# German Submarines and their Warfare 1914-1918

The Great War at Sea Series

Volumes in the Great War at Sea Series

V188 My Torpedo Boat War Missions. Friedrich von Callisen
The Submarine War 1914-1918. Andreas Michelsen
The Last Cruise of the Majestic. George Goodrich
Two Lone Ships. Georg Kopp
U-boats Westward!. Ernst Hashagen
Six Years of Submarine Cruising. Johannes Spiess
German Submarines and their Warfare 1914-1918. Albert Gayer

Cover artwork: Willy Stöwer's representation of U21 stopping S/S Linda Blanche on January 30, 1915.

# German Submarines and their Warfare 1914-1918

Lieutenant-Commander Albert Gayer

Volume I

August 1914 to February 1915

Translated by Celestino Corraliza

Published by Trident Publishing
Miami, Fl USA


**Library of Congress Cataloging-in-Publication Data**

Originally published Die Deutschen U-Boote in ihrer Kriegführung 1914-1918
Berlin 1920
German Submarines and their Warfare 1914-1918
ISBN-13: 978-0-9963157-9-1
World War, 1914-1918 -- Personal narratives, German.
World War, 1914-1918 -- Naval operations.
Military operations, Naval--Submarine.

Printed in the United States of America

# Translator's Forward

Albert Gayer (1880-1930) served as commander of the III Submarine Flotilla, from the war's commencement to late 1917/early 1918 and subsequently to the Armistice in the Imperial Navy Office dealing with submarine construction issues. His wartime operational command and leadership merits are indicated by his sobriquet, "King of the Ems", bestowed by fellow submarine officers under his command. His work was considered indispensable by authors Gibson and Prendergast in their outstanding work chronicling the German submarine war.[1]

He excelled as a field commander during the war pioneering new technologies and supporting boat commanders as the submarine war evolved and matured. Gayer does not directly refer to himself in the text, we see "the Commander of the III Submarine Flotilla ....." as an identifier to an endeavor directed by himself. We have evidence of individual initiative, dispatching all available submarines immediately upon receiving reports of British forces in the German Bight; taking to sea in a surface support vessel to coordinate ship-to-submarine communications as a major fleet engagement appeared imminent; and taking charge of a sunken submarine's salvage to rescue survivors trapped onboard. As with other notable flotilla commanders, Gayer sortied with submarines to acquire firsthand experience of the challenges facing boat commanders in this hitherto novel submarine war. His command ship at the Ems became the center of information exchanges as returning submarine commanders reported updates to British traffic patterns and security measures.

After his tenure as flotilla commander Gayer was posted to the Imperial Naval Office dealing with submarine construction. He touches on the submarine construction issues in the text which would be critical for Germany's prospects in 1917 onwards. He was kept on in the much reduced postwar navy and retired as a Rear Admiral.

Gayer's "Die Deutschen U-Boote in ihrer Kriegführung 1914-1918" encompasses four volumes published in 1920(Vols. I, II, III), 1930(Vol. IV), narrating submarine operations to include notable individual patrols, theater summaries, technology improvements, tactics and frontline personnel views of strategies developed at the most senior military and political level. The implementation of politically inspired strategies of using submarines to stalk warships or merchant ships are described. Although the title indicates coverage through 1918, Gayer did not publish beyond the February 1917 decision authorizing unrestricted submarine warfare. Germany's official history of the submarine war (3 volumes published in 1932, 1933, 1934 cover the same period Gayer does.[2]) did not detail the 1917-1918 phase until 1941(1917 operations), at the time declared confidential with limited circulation and only available to the public in 1966 along with the final volume covering 1918 operations. The presumption is Gayer did not publish further due to the nature of the information which required suppression from public scrutiny, the limited warfare mostly conducted between 1914-1917 being suitable for publication. The format of Gayer's work is similar to Spindler's "Handelskrieg mit U-Booten" official volumes in segregating events by theater and month with an introductory narrative and conclusion to each section. Gayer summarizes patrols and includes noteworthy experiences whereas Spindler chronicles every submarine patrol, occasionally including daily activity.

The work is particularly useful for presenting the views of front line flotilla commanders who are able to analyze their boat commanders' experiences. Their perspective is useful in seeing the reasoning behind submarine deployments. The views of individual boat commanders appear in these pages. (Weddigen, Schwieger, de la Perière) Personnel who commanded support detachments and

---

[1] R.H. Gibson and Maurice Prendergast; "The German Submarine War 1914-1918". London: Constable & Co., 1931.

contributed to the whole are individually recognized. We have an eyewitness account of Admiral Scheer's decision to suspend submarine operations in the North Sea in 1916 which in some respects contradicts the latter's memoirs.[3]

The original text has been entirely translated. There are no omissions; Gayer's footnotes and the foreword for each volume are included. Pagination is now as one complete volume rather than recommencing with each volume. The original volumes did not include maps and illustrations. This version adds cogent items of both to clarify the text and provide visual clues of the separate war theaters, mining operations, submarine types demonstrating the evolution of design, etc. Footnotes have been added correcting misidentification of sunken vessels in the text; the original work was created prior to comprehensive information being made available identifying Entente and Neutral powers ship losses and the Anglo-German information exchanges which amended the official "white papers" used by Gayer.[4]

Gayer did provide an inkling of what would be contained in further volumes in a lecture delivered in 1926.[5] An Afterword has been created summarizing his salient points regarding German Submarines and their Warfare for the 1917-1918 period.

C.Corraliza

---

[2] Arno Spindler, "Der Handelskrieg mit U-Booten", Berlin, E.S. Mittler & Sohn.
[3] Reinhard Scheer, "Deutschlands Hochseeflotte im Weltkrieg" Berlin, Scherl, 1920.

[4] Data presented in tables has not been revised by the editor.

[5] Unable to trace the original, the source used here is a contemporary translation by Commander W.P. Beehler, USN, titled "Summary of German Submarine Operations in the Various Theaters of War From 1914 to 1918", published in the United States Institute Proceedings vol.52, no.4, whole no. 278, April 1926, pp.621-659.

KKpt. Albert Gayer

# Table of Contents

## Part I - From the beginning of the war until February 1915

# Forward

The following narrative, of which this volume is the first, does not represent an attempt to give a complete history of the German submarine war. This remains reserved for later official historiography.

It is very important that the general public, who has heard so much concerning the submarine war from persons little acquainted with the subject, should be told a little more concerning the class of warfare which represented our Navy's concentrated efforts during the war, and in which extraordinary feats were performed.

As the author was Commander of the III Submarine Flotilla from the beginning of the war till January, 1918, and from January, 1918 until the end of the war he was in the Imperial Navy Office's Submarine Bureau, it is only natural that activities in the North Sea will be given prominence in the following pages. For the same reason he is able to deal from an expert's point of view with the question of submarine construction. The narrative will not deal with minor details: it will touch mainly upon those important phases of the submarine war that were decisive.

From this perspective and structure arise the topical divisions: The discussion of the period from the beginning of the war to the start of independent submarine operations introduces the narrative.

The first volume deals with the period of submarine attacks against warships and the beginning of the submarine warfare against hostile mercantile vessels, according to cruiser warfare principles, from the beginning of August, 1914, to the end of February, 1915, and finally the declaration of the submarine blockade.

Four further volumes will follow. The second volume will deal with the submarine blockade from 1915 to the end of the submarine war against merchant vessels in the North Sea; that is, from February to October, 1915.

The third volume covers the period from winter 1915-16 to the beginning of purely submarine cruiser warfare, and the beginning of the exclusive use of submarine against hostile warships in the North Sea — i.e.. October, 1915 to May, 1916. The fourth volume covers the period of submarine cruiser warfare from May, 1916, till February, 1917, and the period during which the North Sea submarines were used against enemy war vessels — i.e., May to September, 1916. The fifth volume deals with the much-discussed unrestricted submarine warfare, from February, 1917, to the end of the war. [6]

It will be seen the referenced volume subdivisions almost inevitably results from operational principles of boat use. The only exception to this is the winter period 1915/16, in which no sharply delineated operative idea is expressed, as the later description will show.

There are a number of reasons which prevent an immediate publication of all five volumes. We are, therefore, forced to limit ourselves for the moment to the first volume.

It is naturally impossible to discuss the submarine war by itself separated entirely from the other happenings on land and sea. We will occasionally be forced to discuss critical points of the war, especially concerning Germany and her allies, as well as the activities of the high sea fleet.

The author has tried to refrain from all criticism now that the war is over. When he criticizes the action taken by the leaders, or where he criticizes important decisions, they will only be criticisms in the light of the knowledge available at the time.

Berlin, January 1920.

[6] As mentioned in the translator's forward Gayer did not complete the fifth volume.

# Submarines against Warships. August 1914 to February 1915

## From the Outbreak of War to the Beginning of Independent Submarine Operations.

At the outbreak of the war all our submarines were at Kiel; boat crews who had just returned there from maneuvers were given leave and rest.

For immediate operations only "U5" to "U25", inclusive, were available. Of these only boats from "U19" onward were fitted with truly reliable engines (Diesels), "U23" and "U25" had up to that date just carried out trial trips and their crews were not fully trained. "U5" to "U18" were fitted with Körting petrol engines. These boats experienced a great number of breakdowns due to ignition troubles. During daylight these boats gave their position away due to their thick exhaust smoke clouds; during darkness the engines misfired and the resulting exhaust flame and noise indicated their position.[7] (see Fig 1. overleaf for profiles.[8])

"U5" to "U25" (at the time, comprising the I and II Submarine Flotilla) departed from Kiel for the North Sea together with the units of the High Sea Fleet that were there in August, 1914; they arrived, with few exceptions, during the evening of August 1st at Helgoland.

At the beginning of the war the hope for the submarine force, as well as the High Seas Fleet, was to enter combat with the enemy as soon as possible. Maneuvers in May, 1911 demonstrated submarines should be very successful in these combined operations, even when

---

7 The Körting engines created an unusual sound similar to a four-stroke cycle gasoline engine backfiring which was audible over long distances.

8 Hans Techel, Der Bau von U-Booten auf der Germaniawerft. Berlin, Verein deutscher Ingenieure Verlag, (1922). Profiles added by the editor in this text are from this work.

pitted against superior forces. The disposition of the High Sea Fleet was made on the assumption that the English fleet would attempt to break into the German Bight. Almost without exception all strategic and tactical training of our submarine flotillas had been based on this assumption. It was only within the last two years that it was realized England would blockade the North Sea by a distant blockade. One event preceding England's declaration of war may be mentioned here. On the evening of August 1st, 1911, the Commander of the High Sea Fleet[9] sent out the signal, "Be in readiness for war against France and Russia; in the meantime avoid at all costs measures against British war and merchant vessels." Though the danger of England joining our opponents did not frighten any submariner, nevertheless the way in which this signal read showed us we were in a politically uncertain situation, quite the opposite from the years 1866 and 1870, when Bismarck had a quite clear idea of the situation.

As an enemy attack was expected, submarines, together with torpedo boats and minesweepers, took up their positions near Helgoland. As the enemy was not inclined to enter the German Bight except with his submarines, the first operations in the North Sea took place on the August 6th. This advance was to ascertain if a blockade really existed; and if so, to learn the position of the blockading ships.

Thus began the first period of independent submarine operations, which lasted from the beginning of August, 1914, until February, 1915. During this time we attempted to weaken the enemy fleet by the submarines who operated mostly without surface fleet support.

Submarine commanders were not very optimistic as to this type of warfare. The means of enticing enemy war vessels near submarine positions seemed to be lacking and as the boats were slow, both submerged and on the surface, it seemed unlikely in most cases that they would be able to get close enough to fire a torpedo even after sighting a ship. Nevertheless, each commander understood the great value of such cruises from a scouting point of view and all crew members were very proud to venture forth alone in these small boats against the enemy with hopes of obtaining a hit with a torpedo. When submarines left harbor or returned their crews,

---

[9] Admiral Friedrich von Ingenohl.

made up almost entirely from volunteers, considered their comrades on the big ships as unfortunates.

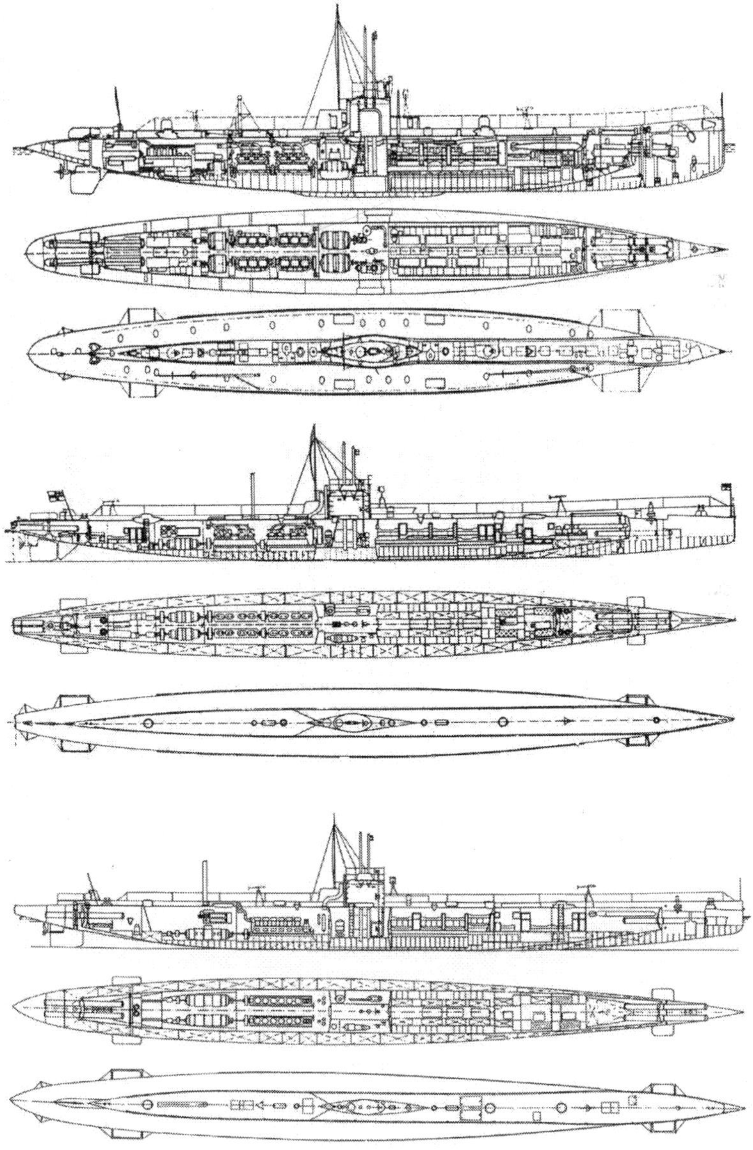

Fig 1. U5-8, U17-18, U19-22 respectively top to bottom.

## The Period of Independent Submarine Operations against Warships and Initial Submarines Operations against Enemy Merchant Ships according to the Principles of Cruiser Warfare.[10]

From the beginning of August, 1914, until the end of February, 1915, with the declaration of the submarine warzone.

At the beginning of this period the German General Staff hoped to finish the war quickly and victoriously, as per plan of Graf von Schlieffen, which was based upon an early decision in the west, employing as many troops there as possible, whilst in the east the defensive was to be assumed until it became possible to employ troops from the west in the east for the decisive battle.

A victory was to be gained by the armies on land, whilst the fleet, with the exception of submarines, was to adopt a waiting attitude. The possibility of weakening or destroying the British fleet by surprise attacks from our numerically smaller but splendidly-trained fleet was not considered. What possibilities were lost only the future can tell. The plan of the high command to win on land only against allied forces, in which England was included, was without doubt a great error. A sea power like England could not be vanquished on the european continent and forced to make peace. Furthermore, at the beginning of the war, when many things concerning naval matters, tactical as well as strategic, were obscure, the weaker fleet undoubtedly had better chances than at any later period of the war; especially so when one considers the tremendous power and material at the disposal of Great Britain, greater than our own by far. We know today from the writings of the British Admiral Jellicoe, who was at that time in command of the Grand Fleet, that during the first month of the war the British fleet was less superior and less ready than at any later date.[11]

This discussion of the general war plan is necessary before considering the activities of submarines under this heading. It will be referred to again later.

---

[10] Cruiser Warfare is synonomous with Prize Rules for the purposes of this text.

[11] Footnote in the original. We know from this English admiral's (then commander in chief of the English fleet) descriptions of the English fleet, its material deficiencies in the war's first months and its readiness for action were less than later in the conflict.

The first submarine sally was made from Helgoland toward a line drawn between Shetland and Bergen. This sortie was carried out in wide formation by the boats “U5” to “U18”, i.e., the old petrol-engine boats. Several of these boats made contact with the enemy. “U15”, commanded by Kapitänleutnant.[12] Pohle, torpedoed and damaged a large British battleship, but shortly after the attack was sunk by the light cruiser “Birmingham”.[13] “U13”, commanded by KptLt. Graf von Schwinitz, was lost, there being no information as to the cause of her loss.

A few days later boats “U19” to “21”, “22”, and “24” were ordered to advance toward a line between Maas Lightship and Outer Gabbard Lightship. “U22” came close to British destroyers on patrol, but nothing else of importance occurred.

These first cruises emphasized the necessity for numerous technical improvements in the boats. When a submarine dived to hide from the enemy, the telephone and lifting buoys, which were safety equipment for peacetime conditions, became detached and rose to the surface. So they were removed. The command bridge rigged in peacetime on the conning tower proved useless for war purposes; a new type of bridge was installed from scratch behind the conning tower. An alarm gear which would call the crew as quickly as possible to their diving stations when the enemy was sighted was also necessary. This was fitted in the form of loud alarm bells. The technical department supervising submarine construction at Kiel studied all these suggestions with the greatest precision and adaptability.

A sally of the cruisers “Strasburg” and “Stralsund” into Hoofden (southern portion of North Sea) was made in the middle of August. They took two submarines with them and gave battle to hostile destroyers, but the submarines had no opportunity to attack.

“U21”, commanded by KptLt. Hersing, carried out at this time a scouting cruise, which traveled past southern Norway to the Moray Firth and the Firth of Forth and enabled us to obtain information

[12] Corresponds to contemporary USN rank of Lieutenant. Hereinafter, KptLt.

[13] “ KptLts. Henning and Siemens reported “U15” damagd a battleship.

concerning enemy coastal patrols composed of torpedoboat destroyers and armed trawlers. The distance traveled of 1600 nautical miles.[14] at that time appeared as a record achievement for a submarine.

The first month of the war came to an end with similar undertakings and cruises by submarines. Besides meeting a few hostile submarines in the German Bight that were scouting and had occasionally fired a torpedo that missed its mark, one noticed nothing of the enemy. This was the reason why the Fleet Command issued a daily order to ships stating that the enemy apparently did not intend to make an entry into the German Bight and he would probably delay doing so until daylight became shorter. The rigid battle requirements demanded for all our ships was accordingly somewhat eased; and some of the larger ships, among them the "Friedrich der Grosse" went into yards to be refitted. The Fleet Command's reasoning proved correct. The enemy finished his scouting through which the position of our advance guards was ascertained, and as his submarines had gained information that our torpedo boats and light cruisers advanced about 25 to 30 miles from the Outer Elbe Light Vessel without support from the larger ships, the enemy deemed it advisable to make a surprise attack with more powerful ships that would cut off our retreat to the mouth of the Elbe and so destroy our advance guard. This plan was carried out on August 28, and was favored by the low visibility.

Should our light forces be sighted by the more powerful enemy, it meant certain destruction; because of the low visibility, they would be unable to sight the enemy until within range of his guns. The sinking of the "Mainz", "Köln" and the "Ariadne" was brought about in this manner, as these vessels, with great bravery, went forth to meet the enemy when torpedo boats had warned them of the enemy approach. All the scouting torpedo boats, except "V187", were able to escape, thanks to thick smoke screens and their high speed. It was very unfortunate that the submarine zone, which during the first weeks of the war was not far behind that of the torpedo boats, had been given up. As at that time submarines were required elsewhere: otherwise one of them would almost certainly have been able to attack the hostile First Battle Cruiser Squadron[15], or light

[14] "Nautical miles" hereafter shortened to nm.
[15] The British Battle Cruiser Squadron providing heavy support for the raiding force.

cruisers. When more powerful German vessels arrived upon the scene the enemy had retired, taking with him prisoners from the sunken German ships.

On this day in Helgoland there were about a dozen submarines ready. They had already received orders to advance westward, but were recalled by the Commander of the Helgoland Submarine Flotilla, as he did not consider that submarines would be successful owing to poor visibility.

August 28th might easily have been a great success for our young fleet if the fleet commander had deduced the conclusion, from the frequent presence of hostile submarines in the Helgoland Bight that these submarines were scouting and preparing for an enemy operation. If for one whole month we had changed our defensive plans accordingly instead of using only light cruisers assisted by torpedo boats, it would have been a very good plan to have had the big ships go out in the evening on all days where the weather was favorable for an enemy entry into the Bight, these ships might have been in an advanced westerly position on the next morning. If the enemy then carried out his attack on the morning of the August 28th, the enemy cruisers would have unexpectedly met German battleships. The battle might have been a great success. The days on which the weather was favorable for attack were very few, and the enemy was forced to bring his torpedo boats as defense against our submarines. The full use of destroyers was only possible when the sea was calm.

The battle of the August 28th depressed the morale of the High Sea Fleet considerably and during the ensuing period all steps possible were taken to make the work of submarines as successful as possible.

To begin with their disposition was changed. It appeared desirable to move some of the boats to the Ems as they were able to proceed north more quickly from there and could in this way protect the flank of the German Bight toward the west. On September 3rd the III Submarine Half-Flotilla, comprising "U19", "U22", "U24, and

"U28", was sent there. The tender of the II Submarine Flotilla[16], the "Stettin" followed soon after.

For a more appropriate management of all submarine forces at the disposal of the High Seas Fleet's commander, Korvetten-kapitän[17] Bauer, who up to this time had been leading the I Submarine Flotilla, was now appointed Commander of Submarines.[18]

The IV Half-Flotilla, was also directed to proceed to the Ems, comprised the new boats "U27", "29", and "30". This flotilla was to finish its training in the Baltic: therefore during the month of September various boats from the Helgoland Flotilla were sent to the Ems as reinforcements as it was thought possible the enemy would repeat his raid into the German Bight.

Submarines sortied repeatedly during this period from the Ems against hostile forces, which had been announced to be north of Ameland and Terschelling by our air scouts, but they were not able to score successes. The enemy of course knew submarines had been moved to the Ems and was correspondingly careful.

It is appropriate to mention the tremendous advantage enjoyed by the British Admiralty due to the neutral States of Holland and Denmark adjoining the German Bight. The enemy thus enjoyed extraordinary facilities for spying and one might say the British Admiralty obtained reports from their agents in the neutral states concerning the disposition of our ships, the weather in the German Bight and other information whenever they contemplated entering the Bight. The knowledge of being under observation in this way was very detrimental to our fleet operations in the German Bight. For this reason our ships did not like to go to the Ems as they feared that their movements would be reported to the enemy.

While submarines were being utilized for the defense of the German Bight, the offensive was not being neglected. "U20" and "U21" were sent to the Firth of Forth at the beginning of September. The commander of "U21", KptLt. Hersing, scored on September 5$^{th}$ at

[16] The III Submarine Half Flotilla was part of the II Submarine Flotilla.

[17] Lieutenant-Commander in USN parlance. Hereafter KKpt.

[18] Bauer continued as Commander of North Sea Submarines until June 1917. He is referred to the text by name and/or title.

1.45 p.m. the first submarine success of the war, yes of history, by torpedoing the small British cruiser "Pathfinder", which was returning to the Firth of Forth. This first success was followed on September 22nd by Weddigen's historical deed in sinking "Hogue" and "Aboukir" and "Cressy". "U9" was detailed to attempt for the first time the passage of the Straits of Dover but not far from the Maas Light Vessel, near the Hook of Holland, the three ships mentioned were sighted. The second and third came to the assistance of the first trying to save men, and all three fell prey to the torpedoes of one submarine.

These three ships were on patrol as in the days of old, proceeding very slowly, and without destroyers to protect their flanks. The orders issued by the British Admiralty shortly after this occurrence did not leave submarines much hope of meeting any more ships behaving so carelessly.[19] But the keenness of submarine commanders was roused.

On October 15th the old English cruiser "Hawke"[20] fell prey in the northern part of the North Sea to another well aimed torpedo by Weddigen.

On October 18th "U27", commanded by KptLt. Wegener, who later fell victim in the "Baralong" atrocity, torpedoed the British submarine "E-3"[21] near the mouth of the Ems. This was the result of a plan made to counter enemy submarine operations at the mouth of the Ems. Ever since German submarines had been stationed at the Ems, English submarines had tried hard to surprise them when entering or leaving the river. The German torpedo boat "S-116" was sunk on the October 6th, close to the mouth of the Ems by a hostile submarine.

During the last days of September "U18", commanded by KptLt. von Henning, passed for the first time through the Straits of Dover. This passage was difficult on account of the strong current and numerous sand banks. This boat found buoys in the same positions

---

[19] References is for large warships to continuously zig-zag and not to stop render assistance to damaged vessels.

[20] Footnote in original. 7,350 tons.

[21] Footnote in original. 791 tons.

as in peacetime and light vessels were showing lights as usual. In the straits small cruisers, destroyers, and submarines were encountered. "U18" attacked the small cruiser "Attentive"[22], which was examining a steamer, but just missed her. The enemy retaliated to this passage by mining the whole section between the Goodwin Sands, Rutingen Sands, and the French coasts. But fortunately, the English knew little about minelaying at that period. The mines were laid mostly on the surface, so that one could avoid them during the day when on the surface; this was possible, as the enemy avoided the mined area.

From this period forward there was more activity in the Straits of Dover. "U19" was ordered near the end of October to get close to Zeebrugge, which had been taken by the Army; but by an unfortunate accident the boat collided with the British destroyer *Badger* on a very dark night and was pushed alongside her. The commander, KptLt. Kolbe, dived and escaped destruction. The enemy assumed that the boat was sunk and published reports accordingly, but this boat worked during the entire period of the war. In February, 1918, when commanded by KptLt. Spiess, she sunk 36,000 tons of mercantile vessels in the Irish Sea.

The first cruises through the Straits of Dover were carried out by "U20", commanded by KptLt. Dresner; "U28", commanded by KptLt. Freiherr von Forstner; "U30", commanded by KptLt. von Rosenberg Grusczni. "U20" returned around Ireland and Scotland and proved that both the material and crew were equal to such a long journey. "U29", commanded by KptLt. Plange, soon after repeated the same trip. "U27", commanded by KptLt. Wegener, sank on the 31st of October, the small British cruiser "Hermes"[23]not far from Calais.

When on November 3rd the High Seas Fleet took the offensive for the first time, the mission being to bombard the English coast close to Yarmouth, all available submarines were placed in such positions that they could report hostile forces or cover the retreat. But as the German cruisers were not disturbed by enemy forces the submarines did not come into action.

---

[22] Footnote in original. 2,750 tons.
[23] Footnote in original. 5,600 tons.

These independent submarine operations took place not only in the south but attempts were also made to sink hostile vessels in the north.

"U16", commanded by the enterprising KptLt. Hansen, penetrated into the harbor of Lerwick in the Shetlands, but did not find any warships there; only merchant vessels.

"U22", commanded by KptLt. Hoppe, searched on two cruises all possible hiding places, in the north of Scotland, especially in the Minch, but did not find any hostile forces.

Full of impatience, "U18", commanded by KptLt. von Hennig, penetrated past the Pentland Skerries into the Hoxa Sound, which is the southern approach to Scapa Flow; because it was certainly expected large British warships would be there. The boat managed to penetrate to a position where it could search the anchorage, which represented a masterpiece of navigation owing to the number of rocks in the channel, but the boat was disappointed, for there were no large ships in the bay so the boat had to return without having scored a success. It was a very hard fate that U18 ran upon the rocks when chased by destroyers and had to be abandoned and sunk by her crew, as she was so badly damaged.

## Initial Observations of the Submarine War and Early Activities against Merchant Vessels.

Despite all efforts made and especially when compared with the number of miles covered and ships lost, the result of the submarine warfare from the beginning of the war until October, 1914, was unimportant.

Even before Weddigen scored his success of sinking warships the question of using submarines against hostile merchant vessels had been discussed with the Commander of Submarines in Wilhelmshaven as it was considered this would be more effective than submarines only stalking warships. This suggestion was not a new one for Conan Doyle had already shown in one of his stories how a small enemy State could bring England to her knees by employing a few submarines capable of sinking merchant ships.[24] Conan Doyle's description was very tempting and it seemed easy in his story but many submarine commanders were anything but optimistic regarding this matter.

During November, 1914, Großadmiral[25] von Tirpitz inquired of the American journalist Wiegand[26] as to what position America would take if Germany should begin a submarine war against hostile merchantmen. In a good many cases members of the submarine arm had the impression the interview's publication would be the signal for England to organize defensive measures against a German submarine war against mercantile vessels and it would have been better to surprise England with the new warfare. But every member of a submarine force was very glad now to belong to the fleet branch which was to be the most active during the war and which was to have a deciding influence on the war, as everybody agreed after the battle of the Marne.

Views differed to a great extent if the boats available at the beginning of 1915 would be sufficient for an effective war against hostile merchant shipping.

The numbers of submarines in actual use had not increased considerably during the period from the war's outbreak to the end of 1914. On January 1st, 1915, there were 20 boats in the North Sea and 7 in the Baltic. "U23" and "U25", which had at first been sent to the North Sea, had been used against Russia since September, 1914 "U26", which was also active in the Baltic, and which was commanded by KptLt. Freiherr von Berckheim, sank the Russian cruiser "Pallada"[27] on October 11, 1914, near the Gulf of Finland. Neither were there many boats under construction, though a special small type that could be built very rapidly had been ordered for the Flemish coast soon after the army had occupied same, but it was considered this type could not be used effectively before June, 1915. The production of big boats was slow and of the 20 boats which

[24] Celebrated writer Sir Arthur Conan Doyle's July 1914 fictional short story "Danger! Being the log of Captain John Sirius" detailing how eight submarines could starve Great Britain into submission.

[25] Fleet Admiral in USN parlance. Hereafter Grand Admiral as the typical translation used in English.

[26] Karl Henry von Wiegand.

[27] Footnote in original. 8,000 tons.

were in the North Sea on January 1st, 1915, 9 were equipped with petrol engines and not very reliable. Under these conditions the opinion of submarine commanders of boats in service was it would be unwise to start with a submarine campaign against merchant shipping on a large scale before the middle of or the end of 1915; and it would be best to prepare for this with all energy by ordering as many submarines as possible, by arming the existing big boats with guns, and in the meantime to continue with combined fleet and submarine operations as had been done in the attack on Yarmouth on November 3rd, 1914. If it could be executed in this manner there was no fear the enemy would develop more effective countermeasures. In the meantime more submarines, one by one, would enter the war against merchantmen; otherwise counter measures might be quite considerable before the submarine campaign would be fully effective. We will see later this view was not taken and the submarine war against merchant shipping was opened on February 18, 1915. The reasons for this decision will be given later.

From the above it is clear submarine commanders accepted at a very early stage the idea of sinking merchant vessels and had begun attacks against such shipping in certain cases since October, 1914. "U17", commanded by KptLt. Feldkirehrer, sank on October 20, 1914, the British steamer "Glitra"[28] according to the principles of the Cruiser Warfare[29] not far from the Norwegian coast, after having given the crew time and opportunity to leave the steamer. It is interesting to note this first sinking of a British steamer by a German submairne was witnessed by a Norwegian torpedo boat which came upon the scene with three cheers for "U17". U21 sank the steamers "Primo"[30] and "Malachite"[31] in a similar manner in November in the eastern part of the Straits of Dover.

The submarine war against merchant shipping became more effective till it culminated in the declaration of the submarine blockade on February 18, 1915. During January, 1915, seven

---

[28] Footnote in original. 860 tons.

[29] Cruiser Warfare, as considered by submarines, is when the attacker allows for safe disembarkation of crew and passengers prior to sinking their vessel.

[30] Footnote in original. 1,366 tons.

[31] Footnote in original. 718 tons.

merchant vessels displacing 17,577 tons gross were sunk; during February nine ships displacing 22,785 tons gross were sunk. Though these results were not decisive, the enemy recognized the necessity for a systematic defense.

## Submarine Activity to February 18, 1915.

At the end of the first year of the war and at the beginning of the second year submarines were used as a means of defense against the blockade of German river mouths, by means of which the British fleet attempted to compel the German fleet to remain in the Jade, Weser, Elbe, and Ems. This was especially the case with boats that had been recently delivered or with those that were not considered efficient enough to make long cruises. These boats were stationed in advance positions off the mouth of the Ems near Helgoland to signal the approach of hostile forces and to attack them if possible. But the boats never saw the enemy except on December 24, 1914, when he tried to disturb German Christmas celebrations by an air attack. On December 25th enemy forces northwest of Helgoland were reported. "U20" was already in an advanced position at the mouth of the Ems. "U22" and "U30" sortied immediately after the report was received. All three vessels sighted the enemy. An attack by "U20" against a small cruiser became impossible as at the very last moment the enemy turned away at full speed. The enemy behaved on this day just as was to be expected according to early submarine experiences. Destroyers changed course all the time, swarmed around the bigger units, and even the latter steered a zigzag course.

## The Beginning of the Flanders Submarine Operations to the Commencement of the Submarine Blockade, February 18, 1915.

We have already stated "U19" had been ordered during the latter part of October to proceed to Zeebrugge, but that she was prevented from reaching harbor.[32] Soon afterwards "U5", "U11", and "U12" were sent there. "U12", on November 12, 1914, sank the torpedo boat "Niger"[33] anchored off Deal. This boat was navigated by KptLt. Fortsmann who became well known later on due to his successes in

[32] Collision with HMS "Badger" referenced earlier in text

[33] Footnote in original. 810 tons.

the Mediterranean and we must concede that this feat was a masterpiece from a nautical point of view as he was compelled to navigate in the shallow water between the British coast and the dangerous Goodwin Sands. Afterwards he tried to sink hostile monitors which were moored between the sand bank close to the French-Belgian coast. He was the first to point out the great difficulties which submarines encountered, which consisted of strong currents and numerous sand banks and very effective enemy defense. "U6", commanded by KptLt. Lemmer, and "U11", commanded by KptLt. von Suchodoletz, were lost during 1915 in Flanders; it is assumed that "U6" ran upon a mine and the "U11" was sunk when attacking Dover.

"U24", commanded by KptLt. Schneider, went to Flanders just before Christmas, operating from the base; this boat sunk during New Year's night, 1914-15, the British battleship "Formidable"[34], which belonged to a squadron of ships cruising off the south coast of England. "U25" had already tried in the afternoon of the previous day to torpedo this ship.

Thus at the beginning of submarines activities in Flanders there appeared those results that were so characteristic of submarines throughout the war, viz, successful attacks and high losses preceded by short cruises. We will see later these conditions determined the future of the submarine operations.

"U8" and "U29" were also in Flanders until February 18.

### The Bombardment of Scarborough and Hartlepool, December 16, 1914, and the Battle of the Dogger Bank on January 24, 1915.

Submarines cooperated in the bombardment of the British coast in December, 1914. The field of operations was reconnoitered very carefully during two cruises by "U27", commanded by KptLt. Wegener, who cleared a path free of mines for those vessels taking part in the bombardment. That submarines did not participate in any other way was much regretted.

---

[34] Footnote in original. 15,000 tons.

First, one had to reckon with the possibility that warships would leave the Firth of Forth and steer south in the case of such a daring attack. Submarines were at the time of the bombardment in such positions to protect the flanks of our ships to the north and south up to a distance of 60 miles from the British coast and would undoubtedly have had opportunities of attack. Proof this theory is correct we will prove later when we discuss the sortie made by Admiral Scheer on August 19, 1916, when in a similar case though no bombardment of a coast took place, two light enemy cruisers were sunk and a battleship badly damaged by submarines similarly placed.

Secondly, if submarines had been used Admiral von Ingenohl might not have been forced to turn with his whole fleet as soon as he did at 8 a.m. on December 16 when at a distance of about 120 miles from the cruisers (who were in the van), and return to the German Bight. He would have felt safely protected by submarines and might have engaged successfully the Second Battleship Squadron[35] which was sighted by our cruisers about noon on December 16 to the westward of Dogger Bank. This squadron was significantly less powerful than our own forces.[36]

In the case of the sortie to Dogger bank (on January 24, 1915) of the battlecruisers "Seydlitz", "Moltke", "Derflinger", the armored cruiser "Blücher" and accompanying small cruisers and torpedoboat flotillas, the presence of submarines would have been very important, as the enterprise was carried out without the assistance of battleships. At the mouth of the Ems there were on the afternoon of the 23d "U19", "U21", "U32", and "U33" at the disposal of the High Command, and these boats could have left on the same day for the rendezvous of our battle cruisers. They could have been used on the 24th during the engagement against the retreating damaged British ships, some of which were being towed. "U21" instead left on the afternoon of the 23d to sink merchant ships in the Irish Sea. "U19" and "U33" were ordered to be in readiness at the Ems and "U32" lay waiting 15 miles north of Borkum. Even in Helgoland several submarines could have been ready for the same purpose. The Commander of Ems Submarine Flotilla, when he

---

[35] Sortied from Scapa Flow consisting of five modern battleships.

[36] Footnote in the original. December 16, 1914 is rightfully reffered to as the fateful day of the German Navy.

received, on the morning of January 24, the message from Admiral Hipper regarding the sighting of hostile forces, immediately sent the three submarines at his disposal to the assistance of our own cruisers, which were engaged in battle with five superior British battle cruisers. But it was too late. The enemy disengaged upon nearing our submarines and mines. Submarines did not have a chance to attack the enemy ships, and Admiral von Ingenohl lost his chance on the very day on which the armored cruiser "Blücher" was sunk. This was the third instance we lost ships due to insufficient support by submarines.[37]

Such days of setbacks were counterbalanced by new deeds by submarines, regarding which the enemy press expressed considerable dismay. "U21", which took no part in the engagement of January 24, left for the Irish Sea, bombarded a hostile airship shed at Barrow-in-Furness and sank three steamers about 15 miles off Liverpool. "U20", commanded by KptLt. Schweiger, at the beginning of February sank three large steamers in the eastern part of the Straits of Dover.[38] "U16", commanded by KptLt. Hansen, and U8, commanded by KptLt. Stoss, also took part in the campaign against merchantmen. "U19" sank the British steamer "Durward" (1,301 tons) and took the crew of this ship to the Maas Light Vessel.

This completes the description of the first phase of the submarine war.

## Conclusion.

We have seen that submarine operations, which were initially directed against warships led to the submarine war against merchantmen. But this change of policy took place very slowly, without attracting much attention, and opportunities to torpedo enemy warships were lost as on January 24, 1915. We will see that in the next period which lasts to October, 1915, enemy counter-measures became very effective- i.e., they had commenced in the period before the submarine blockade was declared. When receiving his orders to commence the submarine blockade at the end of February 1915, an old submarine arm officer remarked that due to

[37] The first instance being the Battle of Helgoland Bight on August 28, 1914. The second is not detailed; may refer to the Battle of Texel on October 17, 1914.

[38] Footnote in original. Total of 9,606 tons.

the losses to be expected and the small number of boats under construction the submarine war against merchantmen in the North Sea would have to cease during the autumn of 1915. We will see this prophecy was correct and the opinion of a portion of submarine commanders was correct when they stated the submarine war against merchantmen should not start before the middle or end of 1915. Otto Weddigen was of this opinion when he left on his last cruise in March, 1915. But why was the submarine war against merchantmen begun so soon? Because a great number of men did not believe the war would last long and were convinced of a speedy victory by the Army. They therefore decided to use the submarines at their disposal as soon as possible, more to assist than to be decisive till it became advisable to make use of the fleet as well. One could arrive at the former conclusion only by feeling instinctively that victory could not be gained on land after the Marne battle; that the surface fleet would only play a secondary role and the decision would be won by a purely submarine war. Then submarine operations should have been delayed for the time when a supply of boats was assured, thus obviating the necessity of periodic pauses in this warfare. The Commander of North Sea Submarines, KKpt. Bauer, was one of the first to clearly recognize that a decision would be achieved via a purely submarine war.

We have mentioned above the number of boats we possessed at the outbreak of war. We will now investigate the gains and losses up to February 1915. In the North Sea the following boats had been added to the fleet: "U27", "U28", "U29", "U30", "U31", "U32", "U33", "U34", "U35", "U36", and "U37". The following had been lost: "U5", "U7", "U11", "U13", "U15", "U18", and "U31".

"U7"[39] was torpedoed by a German submarine, due to incorrect recognition signals. "U31"[40] was lost either because of faulty diving gear or gas poisoning. This boat was found later by the British, drifting, and the interior quite dry, but the crew were dead.[41] How the other boats were lost has already been described above. Thus it is shown only 4 boats were added to the North Sea submarines, making a total of 25 boats. Not more than 25 boats were used on the North Sea front up to the autumn of 1914. Located in the Baltic

[39] Foonote in original. Commander KptLt. König.

[40] Footnote in original. Commadner KptLt. Wachendorff.

[41] Gayer is repeating a wartime myth.

there were at the end of 1914 7 boats which formed the Kurland Flotilla which could for the time being could not be used in the North Sea in the submarine war against merchantmen.

What was under construction at this moment? As it had been recognized submarines could be used advantageously off the coast of Flanders, small torpedo and minelaying boats had been ordered that could be quickly fabricated. These were boats “UB1” to “UB17”, and “UC1” to “UC15”, a total of 33 boats. (See Fig 2 & Fig3) They were so small their components could be carried by rail and then quickly assembled at Pola, or at the yards in Flanders. The decision to build these boats was quite correct, the essential point being to build something quickly, even if the boats were not very powerful. They were fitted with motorboat engines and were, naturally, not of a very efficient type. Furthermore, design immediately proceeded for 94 B.II and C.II submarines, “UB18” to “UB47” and “UC16” to “UC79” — boats which represented an improved type of the small B, etc. Engines for them were ordered so construction could proceed rapidly after the first B and C boats had been tested. The B.II boats were then also quickly ordered built. Construction of the C.II boats was delayed until experience had been gained with the C.I boats.

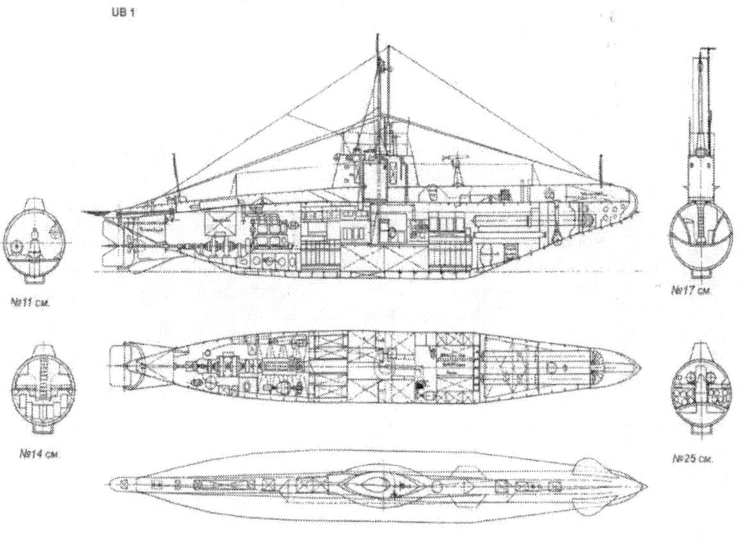


Fig 2.UBI

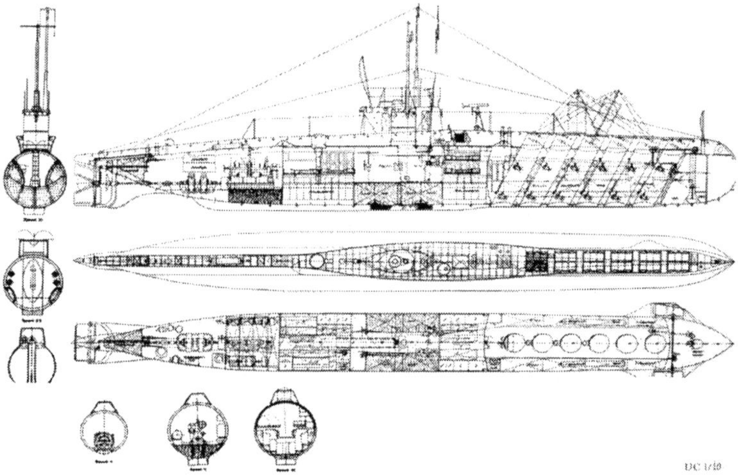

Fig 3. UCI

Ten large minelaying boats had been started at the beginning of 1915, and the following large torpedo submarines had been ordered prior to the outbreak of war, and were either building or undergoing trials: "U38", "U39", "U40", "U41", "U43", "U44" and "U45". At the outbreak of war, "U46" to "U62" had been ordered, and early in 1915, "U63" up to "U65". (see Fig 4, 5, & 6.)

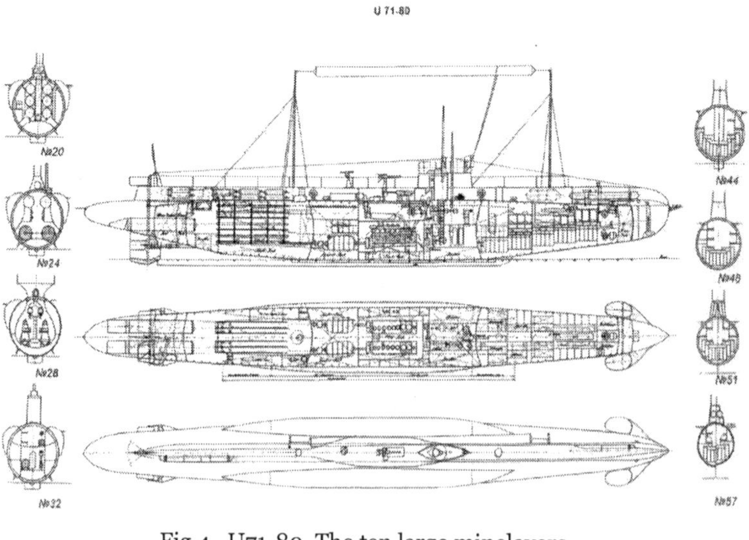


Fig 4. U71-80. The ten large minelayers.

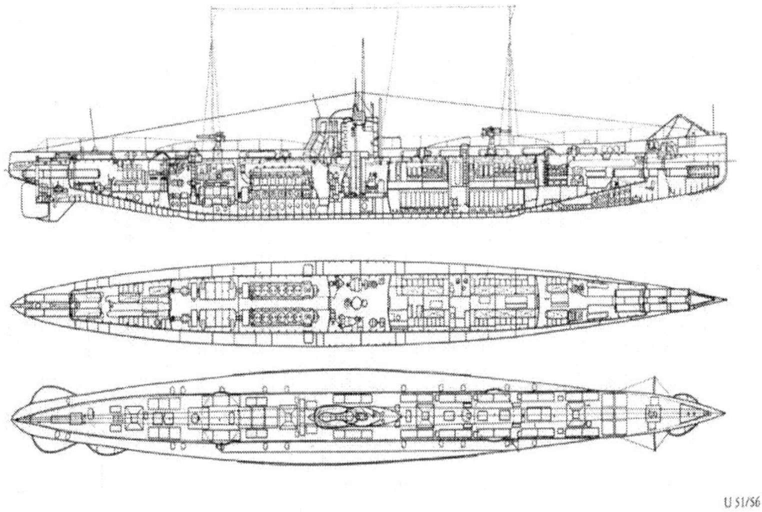


Fig 5. U51-56

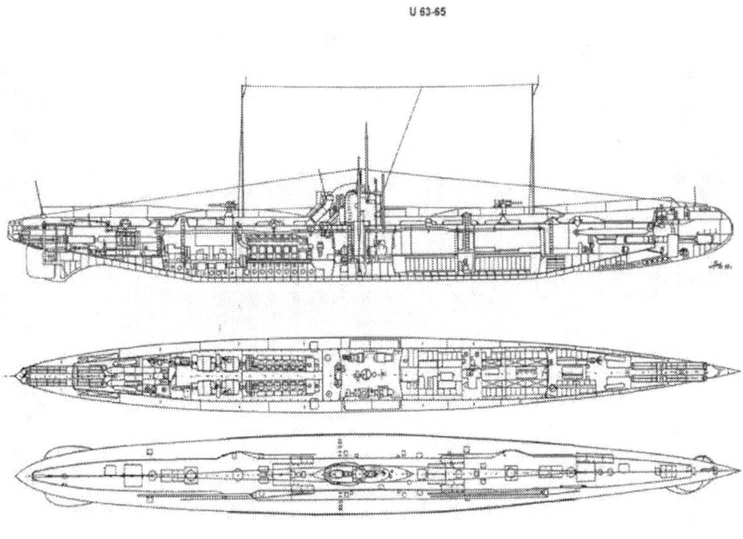


Fig 6. U63 – 65

For the Austrian Navy five large torpedo submarines were under construction, viz, “U66” to “U70”, these boats could be taken over by us after slight alterations.

When the submarine blockade commenced the total of boats ordered and under construction was 42 large and 127 small.

The question naturally arises if this was a large number. We must therefore provide a scale: When the revolution broke out in November, 1918, we had under construction 440 submarines, among them 137 large ones, and a further 330 boats, among them 117 large ones, were just about to be ordered. Of these 770 boats, 360 would have been completed in 1919, and the remaining 410 during 1920, as the Submarine Bureau had laid their plans very carefully.

It would be unfair to state we should have ordered submarines toward the end of 1914, thus conforming to a policy which only became a necessity in 1918. A stage of development was necessary before this was possible, but one might certainly argue that a greater number of torpedo submarines of the type "U31" to "U41"[42] might have been ordered with a few improvements. Then it would have occurred to the department concerned that further yards could be used to build submarines besides the yards then building (Germania, Kiel; Imperial yard, Danzig; Weser yard, Bremen; and Vulkan, Hamburg), one might also have let the inland firms assist in submarine construction, for instance, structural steel firms.[43]

The provision for submarine repairing stations was not undertaken on a sufficiently large scale. This was only done effectively in Flanders (Bruges, Ostend, and Antwerp), due to the Flanders Submarine Flotilla commander, KKpt. Bartenbach. We will see later

[42] The "U31" type was extremely successful in the Mediterranean.

[43] Footnote in original. KKpt. Bartenbach (in the first months of the war still a representative for submarine affairs in the Imperial Navy Office) comments in Grand Admiral Alfred v. Tirpitz's memoirs as follows: "First of all, it was necessary to speed up the readiness of boats already under construction and to reduce to a minimum the time to build newly laid boats, considering the initial 18-month deadline for a boat's construction renders it unable to participate in the current campaign; this rebuke, that is, the rebuke that I did not foresee the duration of the war, I accept." This sentence confirms the correctness of what has been said here, but on the other hand shows that it is easy to reproach today after the whole development is clearly in front of us, while at the beginning of the war it was actually impossible to predict everything correctly. In any case, any criticism that goes beyond that accepted by the KKpt. Bartenbach himself is completely unfounded and without rights.

how very much this point helped Flanders submarine operations during the entire war.

Together with the possibility of constructing and repairing submarines, one had to deal with the question of personnel in a very large way. One had not only to provide submarine crews and train them, but one had also to provide a considerable amount of auxiliary personnel. A large number of engineers had to be taken from the army to assist our naval constructors and had to be trained in submarine construction and repairs. For how else could one provide inspectors and supervisors for the yards and repairing stations? Finally, one had to take or transfer a mass of technicians, mechanics and skilled workers from the army to the navy.

The question arises as to why these activities were not carried out in a timely manner? A number of answers to this question have been published since the end of the war. There can be no doubt that even at the beginning of 1915 Admiral von Tirpitz still hoped to make effective use of the fleet and was not willing to change the entire naval organization to suit a purely submarine program. He had, as can be seen from his memoirs, the idea submarine operations - as initiated by the declaration of a war zone in February 1915 - should only serve as bridge in the time period until the full deployment of the High Seas Fleet. Had he known that this would never occur then he would have changed the whole organization at the beginning of 1915 to conform to a submarine warfare policy.

Taking into account what has been learned from the enemy we may now state decisive success would have occurred in 1917 if the change to submarine warfare had taken place in 1915. But could not the fleet working together with submarines have been successful? We can also answer this question today. A further discussion will show that unfortunately it was never possible during the war to bring our submarines in sufficient numbers to the battle field when our fleet or parts of it gave battle. The reasons for this will be learned later.

In conclusion it can be said here: "At the beginning of 1915, it was less important whether you decided to go to war with submarines or with the High Seas Fleet in cooperation with submarines. The main point was after the setbacks on land, on the Marne, the necessity to strive for a decisive decision at sea was recognized and operationally and materially, sea warfare had at a minimum to be put on an equal footing with land warfare." Grand Admiral von Tirpitz certainly knew this knowledge, and after everything has become known - especially from the enemy - it is certain victory would have been ours if we followed either of the two paths.

In retrospect, it is more than justified to commemorate the pioneering work done by the submarine commanders and crews during the first part of the submarine war.

No matter how great later submarine activity was in terms of brilliant performances, it was more or less based on experiences gained in this initial period.

Submarine crews entered the war very well trained, but there were a great number of precautions in peacetime practice cruises to prevent accidents as much as possible.

With one stroke these precautions were done away with and the boats were compelled to rely upon themselves, far from their bases and menaced by hostile attacks.

Diving in a heavy sea resulted in fairly long dive times in this phase (slow ventilation of the diving tanks), and the boats had to be prepared to dive or fight at any time by day and night. Peacetime training had not sufficed for this practice. During nighttime one had to be surfaced to see what needed to be done. A great help in this case was the invisibility of the boats at night, which was far more pronounced than in the case of torpedo boats. Very often boats passed enemy destroyers quite close during the night without being seen.

It soon became apparent enemy destroyers were a serious opponent to submarines; even if these "cavalry of the seas" were be limited to ramming surfaced or submerged boats for lack of other means in this initial period. Often our submarines tried to attack destroyers — initially especially sought to do by Weddigen — but soon he came to the conclusion that attacks against these "high value targets" was not advisable except under particularly favorable circumstances. Before one is aware, he reports, the destroyer passes over the submarine, and it is difficult to escape the danger by "going deep".

During daytime we had to avoid hostile submarine attacks. The enemy waited at important points on the regular navigational routes to surprise our boats, soon submarines learned to pass those places submerged or surfaced at high speed using zigzag courses.

Soon afterwards we heard of submarine nets, against which one had to provide net deflectors and cutters. The necessary additions were made during the overhaul between trips.

One must also remember the laborious efforts required to keep the oil engines and electrical equipment in order during these long cruises and overhaul periods. The numerous breakdowns of the petrol engines in the older boats strained the engineering staffs to an extraordinary degree. Everywhere engineers, machinist mates and stokers worked day and night to troubleshoot minor or severe problems. How difficult it was to work in the narrow, hot engine rooms of the boats is already known from some publications. Special care in machine operation was of course required for the long first cruises around England. These boats brought valuable experience regarding the best machinery preservation practices under such difficult conditions, laying the groundwork for more advanced operations.

Navigation skills must also be emphasized in their entirety. The difficulties of navigating astronomically from the restless, not very spacious platform were soon overcome and near masterpieces in terrestrial navigation were accomplished. Sands and currents in the English Channel were soon no longer an obstacle, although lying there to wait for a favorable current or night in the area was often uncomfortable for the big boats; for the strong current dragged the boats violently from the bottom; but soon one found favorable places protected by sands or otherwise, where one lay "calmly bedded". There were boats that surfaced through the Minch[44] and the North Channel[45] on bright days or through the English Channel in thick fog and happily reached their destination. (see Fig 7. for North Sea, Hoofden operating areas.[46])

---

[44] Footnote in original. Fairway between the Hebrides and Scotland.

[45] Footnote in original. Northern entry to the Irish Sea.

[46]Fig 7. is from editor's collection.

In view of the common dangers a bond of close feeling, togetherness and trust soon wrapped around leaders and crews. Pride in accomplishments increased eagerness and entrepreneurship, and it was to last four long war years. "Submarine loyalty" was to survive the collapse of our armed forces sea. When the crews of the large ships refused to sail on that fateful October day in 1918 the submarines were outside attacking and hopeful of victory and could not understand developments in home ports.

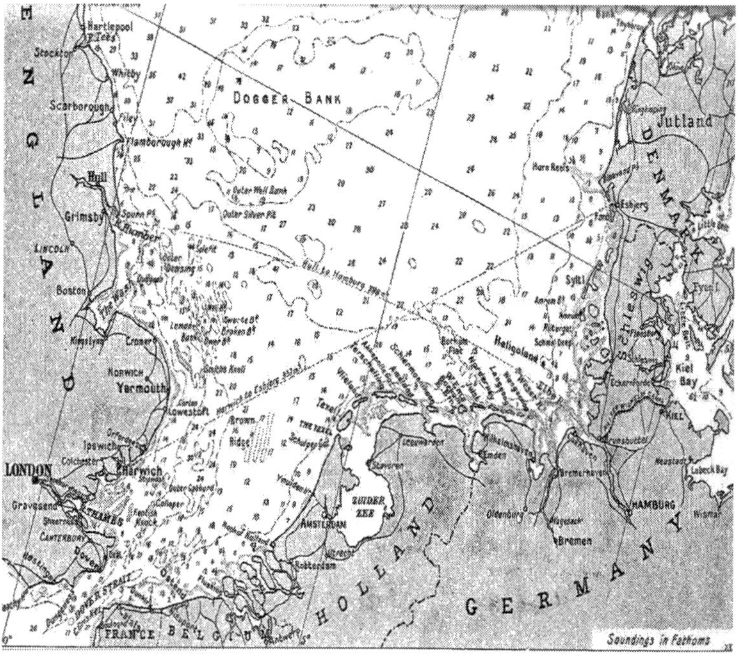


Fig 7. UB/UC & North Sea boat East Coast operations areas.

# German Submarines and their Warfare 1914-1918

Lieutenant-Commander Albert Gayer

## Volume II

## Submarine Warfare Against Commerce February to October 1915

# Part II- Submarine warfare against commerce from February to October 1915

# Forward

Since the first issue was edited, the book of the English admiral "Viscount Jellicoe of Scapa", who was the English fleet commander from the beginning of the war until November 28, 1916, has come out. This book "the Grand Fleet 1914-16, its Creation, Development and Work"[47] has finally given us information regarding many events of the submarine war, of which the veil was only incompletely uncovered. Much of what could otherwise have been speculated could therefore be worked out in the present issue on the basis of Jellicoe's notes as an irrefutable historical fact. The official lists of ship losses during the war, published by the English, Dutch, Swedish, Danish and Norwegian governments, made it possible to supplement and correct information of the sinking results of the boats. Where the sinkings of individual boats are given to the nearest ton, the results are historically flawless.

I would like to express my sincere thanks to all of my comrades who supported me in putting together the events of the war by sharing their personal experiences. At the time this volume was being written, Germany was in a renewed internal struggle. May this text help to recall the time when our homeland was strong through the union of all parties and groups of people. Nowhere was this cohesion expressed more clearly than on the small submarine, where, from the commandant to the youngest fireman, there was only the effort to help the German fatherland win in a joint effort.

Berlin, May 1920 — The Author

[47] Published 1919.

# PART II - Submarine Warfare against Commerce from February to October 1915

## Operational, Organizational and Material preparation Topics

The announcement, made by the German Naval General Staff February 4, 1915 regarding the submarine blockade, read:

1. All the waters surrounding Great Britain and Ireland, including the whole of the English Channel, are hereby declared to be a war zone. From February 18 onwards every enemy merchant vessel found within this war zone will be destroyed without it always being possible to avoid danger to the crews and passengers.

2. Neutral ships will also be exposed to danger in the war zone, as, in view of the misuse of neutral flags ordered on January 31 by the British Government, and owing to unforeseen incidents to which naval warfare is liable, it is impossible to avoid attacks being made on neutral ships in mistake for those of the enemy.

3. Navigation to the north of the Shetlands, in the eastern parts of the North Sea and through a zone at least thirty nautical miles wide along the Dutch coast is not exposed to danger.

This declaration was accompanied by an in-depth memorandum to all allied, neutral and hostile powers, stating that only behavior contrary to international law compelled us to take such sharp countermeasures. The neutrals in particular were asked to keep their people and property away from the war theater.

It is regrettable this explanation, which in itself was flawless, was considerably restricted in its effect at the end of February by further instructions to submarines regarding the protection of the neutrals: the Italian and American flags should be given special protection everywhere. Until the end of March, the Scandinavians had "a free secured strip" for traffic to England. (Fig 8.[48])

Fig 8. Declared Unrestricted Zone.

Subsequent instructions restricting submarine operations follow in the course of the description.

## Material Preparation of the Boats.

The equipping of the boats with artillery - initially only the 8.8 cm cannon came into question - had to be promoted with great haste. With armament still absent on almost all enemy steamers, submarine artillery had gained great importance. It was thus

[48] "The Story of the Great War" Volume V. Edited by Francis J Reynolds, Allen C Churchill, Francis Trevelyan Miller. John A Collier & Son Company, New York, 1919.

possible to stop steamers, against whom the torpedo shot could not be reached due to the great distance, at all distances that were within the range of the submarine's gun. Once the enemy vessel had stopped and the crew was ordered to abandon ship, the choice was made to dispatch the vessel with a torpedo or artillery. We will see how the situation later changed due to defensive arming of enemy merchant ships. Initially submarine artillery came into great effect; especially in war zones such as in the Mediterranean and in the vastness of the Atlantic Ocean, where enemy escorts was hardly available at first. The largest possible amount of ammunition had to be accommodated on the boats. Reconstructions and installations both inside the boat and between the pressure hull and the upper deck were required for this.

In order to keep torpedo consumption as low as possible, large quantities of explosive charges had to be brought on submarines. The explosive cartridge should be used to sink sailing ships and smaller vessels.

## Operational objectives

As far as the operational execution of the new task was concerned, one had to take into account the small number (described in volume 1) and low replenishment of submarines and try to avoid boat losses as much as possible without paralyzing the plan. There is no doubt that this was possible if the boats were instructed appropriately, based on the experience of the most successful commanders. To give an example, the following are instructions that were issued to North Sea submarine commanders partly at the beginning and the remainder in the further course of the submarine blockade:

1. Heavy abuse of the neutral flag is to be expected, because at least all possible protection has been promised. Extreme caution is recommended when approaching neutral vessels on the surface, this should be avoided if possible.

2. It is to be expected that enemy submarines will soon take part in the defense against ours and that they will work with special

"traps", which have the task of attracting our boats. If any of the vehicles already abandoned by the crew are sunk with artillery, the submarines must avoid being stopped. Rather, the bombardment must take place while steering changing courses, etc.

3. Open or concealed guns are to be expected soon on enemy steamers. Against armed steamers the boats are particularly at risk when diving or surfacing. These maneuvers must therefore not be carried out at distances from steamers where it is possible they can fire their arillery with good effect. If a submarine approaches a steamer that appears to have been abandoned by the crew, this must be done under ongoing bombardment in order to force the steamer, if hidden and armed and manned, to show its colors. In the case of reluctance, for whatever reasons, to use a torpedo to sink a ship it is advantageous to carefully inspect the ship while submerged through the periscope for the presence of guns before surfacing to stop it with the cannon and only then surface at a suitable distance from the steamer. If such a previous inspection is not possible, stopping must be carried out by warning shots at distances to which immediate effective firing from the steamer is not possible. For all vessels that are suspicious despite the fact their crews have gotten off, as well as for large and valuable steamers, it is also correct to approach them submerged, after firing artillery to stop them and force the crew off before firing a torpedo.

4. If a submarine has achieved one or more successes in quick succession in a certain area it must relocate its area of activity. How far to relocate will differ in individual cases. As a rule of thumb, movement of around 100 nm is recommended. This relocation is necessary because when submarine activity is reported, which can be provided by most vessels with radio telegraphy before they are sunk, any kind of counteraction will naturally concentrate in the area in question. Above all, it must be expected that enemy submarines will appear after some time. For the same reason, boats are also warned not to remain in the surface for too long in cases of steamers already hit and which do not immediately sink, especially

if they can be seen from being afire; because such vessels are a collective signal for enemy submarine countermeasures.

It can be seen from these "instructions" that the dangers to which the submarines were exposed in 1915, apart from mere submersible attack, were manifold. Well warned, the boats were able to protect themselves against them. When recommending such precautionary measures, the area where the boats were supposed to work played a major role The more submarines worked in a given sea area, the more carefully commanders had to behave; if they came into areas previously only slightly or not touched by submarines, they were able to work much more carefree and exercise caution only under certain circumstances. (An example being in the northern Arctic Ocean in autumn 1916 where several submarines worked.)

These naturally increased the sinking result considerably because the precautionary measures described were mostly associated with a considerable loss of time. Incidentally, it was particularly expressed these "instructions" should not in any way inhibit bold daring of the performance of inventive deeds. On the contrary, such strokes, occasionally made, were all the more likely to be successful than was generally the result of consistent and careful work. Commanders attacked enemy vessels in daytime when the weather was unpredictable or at night, sinking them with quick waterline artillery hits, sometimes submarines intimidated enemy merchant ships by suddenly surfacing in close proximity or by igniting harmless smoke bombs, these surrendered without resistance, but, as mentioned, such a coup could only be afforded by those who were intimately familiar with the warfare, as it was called in the submarine speak, one had to have a "good nose", [for the situation-editor] and even then it had to remain an exception, otherwise there occurred inevitable setbacks.

## Organizational measures for safe patrols

Naturally the risk of boat losses had to be particularly great for new boats entering the front because commanders and crews had to first acquire the necessary war experience. The more enemy countermeasures matured, the greater this danger had to be. How was this to be met? It was immediately decided to let new commanders take part in an instruction cruise with an experienced commandant before taking over their own boat; this measure brought a definite profit, and had the disadvantage if the boat in question were lost, two submarine commanders would be eliminated, a considerable loss in view of the shortage of suitable officers[49]; moreover, an instructive patrol could only rarely exclude the new commander from the extraordinarily varied situations which he was faced with during his own patrols.[50] A better means of providing war experience to new front-line boats was therefore created by the following organization: Within every submarine flotilla, ship officers or navigators from the merchant navy were to be assigned as so-called "war pilots". These war pilots were embarked several times as part of their training on submarines with the most experienced commanders and from the very first voyages they were able to use their knowledge of enemy merchant shipping, especially in distinguishing neutral steamers from enemy ones based on the type, the course used, etc. The more these war pilots sortied, the more they became accustomed to war conditions, so that they were able to support new commanders at the front in assessing many critical situations of submarine warfare. Some of these pilots traveled up to 360 days on submarine patrols and the work of these brave men is also mentioned here. Names like Moll, Rothe, Niejahr, Völcker, Vestmann, Frisch, Frick, Heitmann enjoyed good standing among North Sea submarine commanders.[51] In addition to these war pilots, attempts were made as far as possible to provide radio operators with special war experience to support the actual radio operators on board the boats. Unfortunately, the complete implementation of this measure failed

---

[49] Flotilla commander KptLt. Helmuth Mühlau was captured in June 1915 when on a cruise with "U14" whose commander was killed in the action.

[50] Footnote in the original. It was only possible to promote former submarine watch officers to fill the very few boat commander billets given the strong growth that we had compared to the existing number of available boats.

[51] Footnote in the original. A former captain of the Hamburg South America Line, Reserve KptLt. Crety, made a great contribution to the selection of wartime navigational instructions.

due to the lack of trained personnel, since it did not seem possible to pull enough radio personnel from fleet ships. Radio operators with considerable war experience were also able to provide excellent support to submarine commanders, above all by establishing a secure connection with home stations - even at great distances and by skillfully observing and deciphering enemy F.T.[52] traffic. Especially Senior F.T.-Petty Officers Degen and Heincke and F.T. Seaman Reppert of "U39" have distinguished themselves in this respect.

The boats brought their purely military and diving skills with them from the submarine school, which under the direction of KKpt. Eschenburg developed further during the war period under review here.

As the last and safest means of preparing boats for trouble-free cruises, careful checking and self-tests had to be executed before leaving. If at first there was a certain reluctance of war-bound submarine crews to notice "these drill principles", this soon gave way to the realization of the usefulness of such an initial step. The ever increasing number of personnel changes, the need to shoot older and newer types of torpedo out of the boats, testing the increasing number of technical innovations built into the boats made careful running-in more and more imperative. It was necessary to check submerged boats before each trip to see whether they left an oil trail as this could only too easily betray the boat and lead to destruction with underwater explosives. The fuel oil tanks attached to the outside of the pressure hull of the boats were very difficult to make truly oil-tight and were a constant concern for flotilla commanders who were responsible. [for their condition-ed.].

## Operations Areas

The main operations areas submarine campaigns against enemy commerce were in the area of the English Channel, at the exits of the Irish Sea, on the west coast of France and the access routes to

---

[52] F.T. is an abbreviation for Funken Telegraph(ist) or Wireless. Hereafter F.T.

these waters from the Atlantic Ocean. The main route to these areas - except for the north entrance to the Irish Sea (North Channel) - was through the Dover-Calais narrows. At the beginning of the period we are now describing this route to the operations areas could also be utilized by the large North Sea submarines. Their commanders overcame appalling navigational difficulties with clever procedures; lying on the bottom and awaiting a more favorable current period for steering submerged through the narrows north of Rutingen Sand if a surfaced high speed passage at night was out of the question. Enemy mines hardly bothered at the time, since as mentioned in the first deployment, they were mostly on the surface and they could be avoided during the day, steering surfaced; their ignition system was also not very useful at the time, as Admiral Jellicoe indicates in his book. The return journey through the narrowness from the south offered less navigational difficulties than the passage from the north.

However, a careful observer of the operational situation in that area could not doubt that sooner or later a considerable obstacle blocking the passage with patrols, mine and nets would be established. It was astonishing how hesitantly energetic measures were taken by the enemy in this direction. The Flanders Submarine Flotilla benefited from this for a long time.

The Mediterranean and the Black Sea (Marmara Sea) were added as additional areas of operation in this period of the war.

Speculation whether a large submarine could sortie from the North Sea to the Mediterranean had been occupying the Navy Staff since early 1915. KptLt. Hersing, when asked, agreed to sortie with "U21". We will come back to him in detail. Apart from the initially required effect against the enemy naval forces in front of the Dardanelles there was also a considerable hope the submarine could function in the Mediterranean Sea. It was rightly assumed countermeasures were completely absent for the time being, sea and visibility conditions were considered to be particularly favorable for submarines and it was believed there would be numerous valuable commercial and warlike targets. These prerequisites would also be confirmed by actual experience. In addition to sending submarines by sea to the Mediterranean, war operations there were also prepared by sending nine small B and C-boats by train to Pola. We will see later the first of these boats went into action there simultaneously with "U21".

The Baltic Sea, as in the first period, continued as a secondary theater of submarine warfare.

The boat material: As mentioned in the first volume, blockade activity had to be carried out from North Sea bases by a number of boats that did not exceed 25 simultaneously operational boats in the current war period.

The main burden of the war had to be borne by the boats: “U6”, “U8”, “U9”, “U10”, “U12”, “U14”, “U16”, “U17”, “U19”,”U20”, “U22”, “U23”, “U24, “U25”, “U27”, “U28”, “U29”, “U30”, “U32”, “U33”, “U34”, “U35”, “U36”, “U37”, U38”, “U39”, “U40” and “U41”.

11 of these were sunk by October 1915, they were: "U8", "U12", "U29", "U37", "U14", "U40", "U23", "U36", "U27", "U6", "U41".

Two months after the start of the sbmarine blockade, "U21" was the first boat to travel to the Mediterranean Sea. It was followed at larger intervals thru October 1915 by: "U34", "U35", "U39", "U33", "U38". In addition, six small B-boats and four C-boats which had been assembled in Pola entered the Mediterranean in May and June 1915, respectively. "UB3" was lost in the Mediterranean in the period to October 1915.

The Flanders Submarine Flotilla was created on March 29, 1915 in April with small B and C boats. By October 1915, had increased to 16 boats; the flotilla lost "UB4" and "UC2" in the period to October 1915.

In the Baltic Sea at the Kurland Submarine Flotilla the number of boats fluctuated between five and two boats during the time frame under review. "U26" was lost there.

## February and March 1915

As mentioned in the first volume, in February submarines sank 22,785 tons.[53]

The beginning of the "submarine blockade" was chosen unfavorably as has been referenced earlier. Of the above-mentioned boats that were eligible for warfare, only one ("U8") was ready for departure on the day set for the start of the blockade; another boat, "U30", was already enroute, so on February 20, outside Liverpool, it was able to do fine work with the destruction of the steamer "Cambank" (3,112 tons). The first boats to leave the Ems to execute the blockade operational order were - on February 25, 1915 - "U20" and "U27" under the KptLts. Schwieger and Wegener. Wegener went north around England — Ireland to the west coast and was also supposed to work in the Irish Sea, north of the Isle of Man. Schwieger, who broke through the Channel, was supposed to be active in the Irish Sea, south of the Isle of Man, in the St. George and Bristol Channel.

It was foreseeable that good and secure radio communication with home would be of paramount importance for the extensive boat operations brought about by the new warfare. Ship radio stations in Wilhelmshaven appeared to be unsuitable to function as control stations for submarine operations, as they were too disturbed by the local traffic of the High Seas Fleet. The "Arkona" had always been an excellent F.T. ship; now it is recognized - without external prompting – her communications officer, Oberleutnant zur See[54] Schaafhausen, supported by his excellent staff, that "Arkona" was

[53] Footnote in Original. Mine and warship successes are not included in this tonnage number as in they are in later months tonnage. It is likely that the figures will have to be corrected later by further historical research, as far as they concern monthly results.

[54] Lieutenant Junior Grade in USN parlance. Hereafter Oblt."zur See" designation indicating line officer progression.

far better at calling submarines and could perceive boat messages earlier than other stations. On February 26, an appropriate range test was carried out for the first time with "U27" transmitting. A perfect connection was achieved up to 140 nm.[55] This was done on the initiative of KptLt. Prause, commander of the IV Submarine Half-Flotilla, to which "U27" belonged.

In the first days of March we were under the impression of an attempted English attack and blocking attempt. Everything was ready, but nothing happened. March 1st arrived, "U23", Oblt. Schulteß, the first of the boats which had previously been active in the Baltic, arrived to strengthen the trade war in the North Sea.

At the beginning of March, the first boats came to the Front with a newly installed quick diving device. The diving times were reduced from five to two minutes.

At the beginning of the submarine blockade, "U8" under KptLt. Stoß sailed from Helgoland through the Dover Strait to the English Channel. The boat was successfully engaged in attacks, went to Zeebrugge after it had fired all torpedoes, reloaded and went to work again. Again some successes were achieved (15,049 tons - five steamers sunk), on March 4th it was unfortunate that "U8" not far from Dover, submerged, apparently in very smooth weather, was discovered by an oil trail trailing the boat such that English destroyers were able to use an submerged explosives weapon to attack "U8" and finally force it to surface. The boat sank; the commander and the entire crew came into English captivity. It was now clear what rage against our submarine activity had accumulated in the enemy. The British Admiralty, yielding to public opinion, denied the prisoners the honorable treatment which they are entitled to under international law; they were dragged to prison, one that was otherwise only intended for serious English criminals, mutineers and the like. They were locked up in individual cells. It

55 Footnote in original. Later "Arkona" and submarines were later equipped with new sound amplifiers, the ranges increased to 1000 nm and more. The flag lieutenant of the High Seas Fleet, KKpt. Bindseil, had a large share in this performance development.

was only after 3-1/2 months, through counter-reprisals taken by the German government, before the brave submariners were freed from this sad lot.

In the first days of March, "U12" was under a new commander, KptLt. Kratzsch (KptLt. Forstmann, whom we had previously known as commander of "U12" had taken over the new boat "U39", on a foray off the east coast of England between Peterhead and St. Abbs Head. On March 9, the English steamer "Tangistan" (3,738 tons) was sunk. On March 10, the boat sighted a destroyer in the low visibility morning, dived and attacked. In this attack, the boat was rammed by the destroyer "Ariel" and was severely damaged by an explosion that it had to surface. It was immediately shot at by three enemy destroyers, where the brave commander was killed. Since the condition of the boat was hopeless, the crew had to abandon and scuttle it. War pilot Völcker, who was rescued, was able to escape from the prison camp after a few months and, because he spoke "dialect English", to find his way to the coast after frequent brushes with the English. There he "got" a small sailing boat and took a trip with it across the North Sea. On this he trip he met (September 1915) by chance "U16", which was then operating under KptLt. Hillebrand in the North Sea and took submarine to Helgoland. He brought the report about the sinking of "U12" with him. The use of an explosive device, which was towed by the destroyer at the stern, had also played a role for in her loss. Those saved by "U12" met the fate of the "U8" crew.

The sinking of the auxiliary cruiser "Bayano" (5,948 gross tons) on March 11 was again a success for "U27" under KptLt. Wegener. It took place in front of the Firth of Clyde (at the northern exit of the Irish Sea) in addition to a commercial steamer. Wegener's companion, KptLt. Schwieger, sank three steamers (9,606 tons) in the southern Irish Sea at the same time. At the same time as Schwieger and Wegener, "U23" left the Ems under Oblt. Schultheß on March 17; he sank two English steamers on the east coast off the mouth of the Tyne.

On March 18, "U28" under KptLt. Frhrn. von Forstner brought two Dutch steamers, which had voyaged from England, to Zeebrugge as prizes. In addition, the boat sank the English steamer "Leeuwarden"[56] at Maas Lightship. This was the start of a very successful undertaking by this boat, in which 6 large steamers (20,987 tons) were sunk off the Cornwall coast. The boat worked almost exclusively with its 8.8 cm quick-firing cannon.

At the beginning of March, "U29" commanded by Otto Weddigen sortied from the Ems through Dover Strait to conduct war on trade. Basing at Ostend had been selected for a short time. Soon English newspapers reported on Weddigen's significant successes in the trade war; numerous steamers[57] and sailing vessels had been sunk by him after the rescue of their crews - as English newspapers expressly emphasized. Then there was a great silence - Weddigen did not return – legends were created regarding his end - how he had been brought down - one suspected hostile deceit by misusing neutral flags. The news came later that "U29" had been destroyed in an attack on the English fleet. Admiral Jellicoe finally revealed in his book: "The Grand Fleet 1914-1916". Weddigen did not return through the Channel, but returned north around England and met the English fleet between Peterhead and Norway on March 18. From Jellicoe (Page 208 ff.) First Weddigen attacked the 1st English Battleship Squadron, which steered with 15 knot zigzag turns. Weddigen's torpedo passed close to the stern of the dreadnought "Neptune". The English fleet commander then had the entire fleet turned 12 points to starboard by "turning at the same time" and increased the speed to 17 knots in order to depart from the point at which the submarine was sighted south of the fleet, which had thus turned north and was just passing the 4th Battleship Squadron heading for the Firth of Moray - this squadron had already been detached and its ships had therefore not turned north. So they just arrived at Weddigen's position. "Dreadnought" (the oldest English large battleship) sighted the periscope only 1 point on the port side,

[56] Footnote in Original. 990tons

[57]Footnote in Original. 3 English, 1 French steamers of 12,934 tons are now officially confirmed.

and the commander immediately ordered full speed towards the sighted position. The ramming succeeding, "U29" surfaced by the bows bow so that the enemy could read the boat number painted there. Apparently the boat was badly hit aft because it sank immediately. This episode was kept secret at the instigation of the English fleet commander in order to keep the enemy in the dark about English methods of submarine defense, which, according to Jellicoe, must have had an adverse effect on the morale of our submarine crews.

Let us examine Otto Weddigen's end.

Weddigen had already sunk four enemy warships and now he had the longed-for opportunity to attack the most modern units of the enemy fleet. We cannot ignore this point. It was astonishing how Weddigen repeatedly found enemy warships. Outstanding submarine commanders such as the commander of "U20", KptLt. Schwieger, patrolled for years without seeing a warship, others, such as Weddigen, Hersing, Wegener, encountered warships frequently. The attack was certainly made more difficult for Weddigen due to the enemy's zigzag. It should not be overlooked that Weddigen was returning from a very successful trade war cruise; his torpedo supply could therefore be very limited. For this reason, he had probably had to do without the more promising double shot against "Neptune". Admiral Jellicoe attributes the collision to Weddigen becoming confused by the zigzag and twisting movements of the English ships and in trying to "get clear"and overlooked "Dreadnought" so that it was able to execute the ramming blow in a skilful maneuver. We disagree here. If Weddigen had only wanted to "get clear", he only had to go to a greater depth and everything was safe. No, the collision is only due to the fact that Weddigen knew only one solution, the attack. From the English description "Dreadnought" with on a western course, "U29" with a southern course sighted 1 point on the port side; it is clear Weddigen had apparently launched a stern attack against "Dreadnought" because there were no bow torpedoes left. Our peacetime exercises showed the danger of collision with a submarine is very high when the stern approach is used in the event the submarine is sighted. Whether a certain amount of submerged unwieldiness which was inherent in the boats "U27" to "U30" contributed to the loss escapes our judgment.

From Helgoland at the beginning of March "U34" under KptLt. Rücker, with the commander of the II Submarine Half-Flotilla, KKpt. Spindler, and "U35" under KptLt. Kophamel went through the Dover Straits in the English Channel. Both boats worked there happily and successfully - 4 English, 1 French steamer (13,088 tons) were sunk by them. These boats, which were at the time the very latest type, were particularly suited to cope with passage through the Straits of Dover, thanks to its good surface speed (15 to 16 knots). "U35" covered the main section of the narrow passage underwater and was not affected by any obstacles. At the end of March both boats returned to Helgoland safely.

A different fate befell "U37" under KptLt. Erich Wilcke. This boat left Helgoland on March 20 on the same route as "U34" and "U35". It did not return, and no one fell into enemy hands, so there was np accurate news of the boat's fate. According to rumors, after several successes, "U37" had been rammed and sunk by a steamer in the eastern part of the channel while underwater or when diving.[58] KptLt. Wilcke had been Flag Lieutenant to the I Submarine Flotilla's commander until the end of 1914 and as such had made a great contribution to the introduction of the boats to warfare under the direction of KKpt. Bauer.

The month of March showed a sinking result of 29 ships with 89,517 tons of ship space according to the reports received. The actual result was probably a little higher because some of the sinkings achieved by the lost boats are not included in the above figure.

[58] Contemporary research attributes the loss to a mine.

## April 1915

“U33”, which began its operations in late March on the trade routes between the Netherlands and England, first experienced how apparently peaceful commercial steamers had acquired considerable skills in ramming German submarines. "U33" narrowly escaped being rammed by the English steamer "Brussels", whose captain Fryatt believing the boat was destroyed, reported it and received a gold watch from the British Admiralty as a reward. "U33" found mine fields had increased considerably on his journey through "Straits of Dover". Not only between Rutingen and Sandettie Bank, but also in the western passage between Sandettie and the Falls there were now numerous mine fields, but here, too, the mines were mostly on the surface. “U33” managed to pass without much difficulty. Two steamers and two sailing vessels were his prey in the eastern part of the English Channel. "U32" (Commander KptLt. Spiegel Frhr. v. Peckelsheim) got into a net during a submerged run through the Straits of Dover in early April. The boat had to go to the bottom and wait for darkness to proceed. Surfacing, there was strong net wrapped around the front of the conning tower and extending far aft on both sides. "U24", which made passage on this route at the same time but was proceeding surfaced at night, was able to get through through and back without hindrance. "U32" chose the way back north around England and his report regarding the Dover-Calais net obstruction resulted in the northern route being made mandatory. Since it was apparently possible to escape the underwater nets by passing through at night, the instruction for all North Sea submarines to now avoid Dover Strait seemed too excessive for some frontline submarine commanders. However, the Commander North Sea Submarines believed it was operationally correct to completely abandon Dover-Calais and it was a lesson learned by events which followed later, and we shall see later that a departure from this principle, which KKpt. Bauer had correctly recognized at the beginning of April 1915, later only brought unnecessary agitation and losses; it was in April 1915 the small B and C boats were gathered in Zeebrugge to form the Flanders Submarine Flotilla and began to work in the Channel, and if the North Sea submarines were to work that area at the same time as the small Channel boats it would needlessly concentrate anti-submarine measures there, apart from the greater danger to which large boats in this areas were exposed to a compared to the small ones.

"U22" (KptLt. Hoppe) sailed April 15th rounding the north of England to arrive on the west coast. West of the Hebrides, the boat lost bow trim when diving to attack a steamer in heavy seas; various damages occurred which forced the boat to turn return early.[59] Nevertheless, she still managed to sink two ships. Returning to the Elms on April 23.

On April 20, after sinking two steamers on the east coast of England, "U6" - KptLt. Lepsius - arrived with an English fishing trawler as a prize taken in the Lister Deep.

From April 17 to May 4, "U38" - KptLt. Max Valentiner - was on his first commerce war cruise to the east coast of England. Two sailing ships were sunk, three steamers brought in and taken to Lister Deep.

The month of April saw 40,639 tons of shipping sunk. The newly launched Flanders boats contributed 9713 tons to this total.

The first shot of the Flanders Submarine Flotilla was fired on April 10 at the North Hinder Lightship. "UB4" (Oblt. Groß) sank a steamer there with a torpedo. "UB5" (Oblt. Smiths) and "UB10" (Oblt. Steinbrink) soon sank another steamer in this area. "UB4" followed on April 17 with a second success. Further successes were thwarted in April by various technical malfunctions on the new small boats.

The overall decline compared to March was due to the losses that kept the number of boats in the North Sea the same (23) despite additions and after the first arrivals in March there were now a

---

[59] The battery overflowed during an uncontrolled steep dive, damaging numerous cells resulting in limited submerged travel range.

relatively large number of boats at the shipyards. Of the 17 North Sea boats, the main contributors, in addition to those mentioned in the description, were: "U24" under KptLt. Schneider, who sank four steamers in the western part of the English Channel, and "U28" (KptLt. Frhr. v. Forstner), whose cruise is described on page 47.

## The dispatch of "U21" (KptLt. Hersing) to the Mediterranean Sea

At the end of March, "U21" had arrived at the Ems where the boat finalized preparations for departure after completing the modifications necessary for the Mediterranean trip. This dragged on until mid-April. Many arrangements were still needed to be made by the Admiralty Naval Staff. It did not seem advisable to let the boat leave without the possibility of replenishing stocks enroute and therefore steps were taken to enable her to obtain fuel half way. On April 25th the boat went to sea; making its way through the Silver Pit and the Fair Island Passage north of the Orkneys, at the designated rendezvous point she met the steamer to provide fuel, oil, and other necessary materials. But then the sad discovery was made that the fuel oil was unusable for the M.A.N.[60] engines because the composition was not the required type.[61] Hersing was faced with the difficult choice of whether to turn around or keep to his task. Calculations indicated existing fuel could just be enough to reach Cattaro but only if weather conditions were favorable. The way through the English Channel to home base was still somewhat closer, but it was feared progress would be slower due to bad weather and hostile counter measures. Hersing decided to press on to Cattaro.[62] The Straits of Gibraltar emerged at dawn when the guarding English destroyers had just steered northward, presumably to hand over security to their replacement. On May 13,

[60] Footnote in original. MAN= Maschinenfabrik Augsburg-Nürnberg.

[61] The supply ship carried crude oil instead of the required Diesel oil.

[62] Footnote in original. It must be remembered that Italy's attitude at that time was already so dubious that it would not be possible to expect Italian ports to call in the event of an emergency. (The Italian declaration of war on Austria-Hungary fell on May 2, 1915).

the boat arrived in Cattaro with half a ton of fuel oil left, where - as in Pola - a German "Special Naval Command", based at Austro-Hungarian facilities, already created to some extent facilities for submarine repair and equipment. The "Special Command" was well led by KptLt. Adam, who previously commanded the Baltic Sea Submarine Flotilla, Naval Architect Weichardt and Naval Engineer Dreykorn.[63]

Hersing left Cattaro seven days after his arrival and headed for the Dardanelles through Cerigo Street and the tangle of Aegean Sea islands. He did not attack the Russian cruiser "Askold" (which had come from the East through the Suez Canal to the Mediterranean) at Dedeagach, he wanted to hit the main enemy. On May 25 he sank the English battleship "Triumph"[64] in front of Ari Burnn, and the English battleship "Majestic"[65] on May 27 in the early morning in front of Seddul Bahr. On June 5, Hersing ran into the Dardanelles, greeted stormily in Constantinople as its savior.

With Hersing's success, which was also thanks to the good endurance of the boat's engines, special attention must be given to his senior engineer, Marine Engineer Heine. The Flotilla Engineer of the III Submarine Half-Flotilla, Naval Engineer Papenberg, earned a great reputation; he was one of the first marine engineers

---

[63]Footnote in original. The small boats were already under the direction of this special command (see volume I) "UB1", "UB3", "UB7", "UB8", "UB14", "UB15", "UC12", "UC13", "UC14", "UC15". "UB3", under Oblt. Siegfried Schmidt, "UB8" under Oblt. v. Voigt and "UB7" under Oblt. Werner had already started the trip to the Dardanelles before Hersing. "UB3" sent his last report 80 nm before Smyrna, then nothing was heard from him; the boat probably sank as a result of a diving failure. "UB7" arrived in Ienikale on May 31st after a failed attack on an English destroyer, "UB8" on June 4th at Constantinople, after Oblt. Voigt had struck with a torpedo a large enemy warship with three funnels, but it did not sink. (translator's note:this was a dummy warship, a steamer converted to appear as the English battlecruiser "Tiger". The vessel eventually sank from damage incurred.) "UC13", "UC14" and "UC15" also went to Constantinople during the summer under Oblts. Kirchner, Bauer and v. Dewitz respectively. They brought weapons and ammunition to the Turks, which was extremely important at the time as it was hardly possible to carry these things through Romania. Dewitz was only able to make the crossing by overcoming unspeakable difficulties, because he suffered a serious machinery malfunction on the way. In the bay of Kos (based at Orak near Budrum) he had to wait for a few weeks for help until he was able to continue to Constantinople with repaired engines.

[64] Footnote in Original. 11,985 tons

to work in the submarine force and was indefatigable in the technical development of the new weapon and in training engineering personnel. The outstanding performance of boats "17 19, 20, 21, 22, 24" in the first part of the war was primarily thanks to his work.

Hersing and his team had solved the problem of traveling to the Mediterranean Sea, and in turn had done pioneering work. For the other boats that went into the Mediterranean Sea thru autumn 1915, "U34", "U35", "U33", "U39", "U38", the experience of "U21" were put to good use.

In contrast to the operations of "U21", the activity of these boats sent to the Mediterranean Sea immediately was one of purely a war on commerce, whereby, as we will see later, almost all of the boats mentioned achieved extraordinary success.

We return to the home theaters of war.

At the end of April, U25 under KptLt. Wünsche arrived at Borkum-Reede from the Baltic Sea, (previously Kurland Submarine Flotilla). In Wünsche, an officer entered the North Sea war scene who had extraordinary successes and who was one of the few commanders who were able to persevere in successful work from the beginning of to the end of the war. It was a tough fate that Wünsche succumbed to an accident at his home in Kiel after his war duties were concluded. (March 1919).

## May 1915

At the beginning of the month an event occurred which caused great political upheaval and disrupted our submarine operations. There were multiple reports the giant steamer "Lusitania"[66] of the Cunard Line was equipped as an auxiliary cruiser and brought large amounts of ammunition from America to England with every

---

[65] Footnote in original. 14,900 tons
[66] Footnote in original. 30,396 tons.

voyage. It was clear "Lusitania" had visited a shipyard at the end of 1914 and refitted as an auxiliary cruiser. Submarine commanders knew the German ambassador to Washington, Count Bernstorff, had warned American citizens against the use of the "Lusitania" as a means of transport to Europe. They saw this as confirmation that "Lusitania" was an auxiliary cruiser serving England's ammunition supply.

In any case, when "U20" left Borkum-Reede at dawn on April 30 under her tried and tested commander, KptLt. Schwieger, no one suspected that he was destined to sink the ocean liner and thus be the tragic unwitting culprit of political implications from whose entanglement the submarine trade war could not be released until the American declaration of war in spring 1917.

On May 7th, between 2 and 3 p.m., the weather was fine and clear on the south coast of Ireland near Old Head of Kinsale, when Schwieger sighted so many masts and funnels in front of him he first thought he was seeing a well equipped destroyer flotilla in line-ahead formation. It soon became apparent that "everything belonged to a steamer". It was "Lusitania", however, the commander was only able to identify it after the shot had been taken.

Although the ship sinking right away was doubtful at first, Schwieger was unable to fire a second, ready to launch, torpedo when he saw the large number of passengers attempting to save themselves. As if in a premonition of the tragedy behind his success, "moved by a heavy foreboding"[67], he took the boat to a depth of 20 meters. The exchange of notes caused by the sinking is well known. The operative consequence was the order to the submarines, issued on June 6th, to stop attacking large passenger steamers, even enemy ones.

---

[67] Footnote in original. Schwieger's total score on this trip was: 4 steamers, 1 sailing vessel, a total of 40,027 tons.

On May 2nd, "U26" transferring from the Kurland Subamarine Flotilla arrived at Borkum-Reede under the brave KptLt. Frhrn. Von Berckheim (destroyer of "Pallada", see volume 1). He was thrilled to be able to take part in this important submarine operation in the North Sea. However, this fine hope did not come about; as soon as the boat was trained and equipped for its new task, it was brought back to the Baltic Sea, where the sinking of the Russian minelayer "Jenissei" (June 4, 1915) soon gave him a great success. It was a hard luck that Frhr. v. Berckheim did not return from the subsequent cruise - August 1915 - including boat and crew; "U26" had probably gotten into a Russian minefield. v. Berckheim was one of those commanders who were expected to be a great success in submarine warfare.

The English destroyer "Recruit"[68] fell victim on May 1 at the Galloper lightship to the small submarine "UB6", commander Oblt. Häcker. The Flanders Submarine Flotilla thus booked its first warship success. Further sinking successes were not had in May because almost everywhere in the Hoofden, the preliminary operating area of Flanders boats, only a neutral flag was found, and neutral vessels were not allowed to be sunk without warning and investigation. Because of the great danger to small boats when inspecting steamers, the Commander of the Flanders Submarine Flotilla had correctly prohibited such a procedure. In contrast, very important preparatory work was carried out for the use of the small mine boats (UC boats) to be expected in Flanders. The navigational situation south of the Downs was clarified, and the Dover-Calais barrier and its passage gaps were determined as far as possible. It was thus possible for the first mine boat "UC11" (commander Oblt. Walter Schmidt) arriving in Flanders could be immediately dispatched to a successful mine operation at South Goodwin Light Ship.

"U23" (Oblt. Schultheß) found in the period from May 10th to 25th off the English coast that numerous areas, which the enemy

---

[68] Footnote in original. 385 tons.

declared filled with mines, did not actually exist; Mined areas have numerous steamers of all sizes cruising and also many sailing ships, so the announcement was apparently only made to deter us from coastal bombardments. Two cargo steamers and three fishing trawlers were sunk on this trip. Numerous neutral vessels had to be released again after investigation due to the restrictive submarine war provisions.

Captain Forstmann, previously mentioned in the first volume as commander of "U12", was active in the North Sea with his new boat "U39" from April 29 to May 12. He patrolled from Helgoland via Hauftholm to the Firth of Forth, then north to the latitude of the Pentland Firth and back to Horn's Reef. He sank two steamers, two sailing vessels, five fishing trawlers on this cruise and brought home a prize. On May 10th he sighted 8 English battleships and 4 armored cruisers about 100 nm northeast of the Firth of Forth, on course steering to the Firth of Forth with a 15 knot zigzag course. A torpedo was launched at the trailing battleship of one of the battleship divisions; the torpedo missed its target as a result of a course change the ships made at the moment of the shot. The groups were not escorted by destroyers. Jellicoe's book (page 218) shows that it was the 3rd Battleship and 3rd Cruiser squadron against which Forstmann had attacked and the battleship "Dominion" narrowly avoided the attack. On this cruiser "U39" achieved a record radio transmission range of 340 nm.

Simultaneously in the North Sea worked "U16", "U9" and "U36".

In "U16" Oblt. Hillebrand, who had taken over this boat at the beginning of March from KptLt. Hansen, sank 2 steamers and 2 sailing vessels. “U9" under Weddigen's successor, Oblt. Spieß, destroyed 10 enemy fishing trawlers and 2 cargo steamers totaling about 6,000 tons.

"U36", KptLt. Graeff, brought two prizes to List after sinking a Danish steamer on May 11th.

At that time boats were given new, small-diameter end tip periscopes so they could attack as unseen as possible. Until then periscopes had been of the same diameter from top to bottom. The innovation proved to be a great relief during underwater attack. Soon after the war began it had been found in the example of enemy submarines, very short periscope cross sectional diameters. The question of reliable signaling by submerged boats for correct recognition by surface vessels played an important role at that time. All sorts of attempts were made in this regard at Borkum-Reede, finally achieving satisfactory results.

On May 17, two submarines, "U19" and "U25", with Flotilla Leader boat "G137" and Flag Cruiser "Hamburg" were ordered to take part in a High Seas Fleet sortie in the North Sea aimed at laying a minefield. The submarines were positioned so they stood between the mentioned battlecruisers and the main enemy forces, at a place most favorable to submarine interception when enemy armed forces appeared. If the fleet had been brought to the enemy coast instead of only the middle of the North Sea and about a dozen submarines had been used, a great success could be expected. After the addition of all the "König" class battleships and completion of several small cruisers and numerous torpedo boats, the fleet was at an excellent level of effectiveness at the time. When on the morning of May 18 our mighty 3 squadrons, in the bright shimmer of the season, passed the position of the submarines, on a course turning back to the Jade, the heart sank in the sensation of what we failed to do, by not uninhibitedly throwing these magnificent ships and their excellent crews at the enemy. The fleet command under Admiral v. Pohl[69] was believed to hold similar feelings; May 30th saw the fleet again at sea, heading for the Silver Pit where some submarines were ready to work with the fleet. A storm from the west, which came up in the morning and made the enterprise hopeless, caused the fleet to come back home.

---

[69] Admiral Hugo von Pohl had assumed command of the High Seas fleet on February 2, 1915.

Naval operations in 1915 mostly aimed to lay mines in the open North Sea. These minefields have not brought us any particular success, since the areas in which they were sewn mostly served as the grounds of many Dutch or English fishermen. Fishermen very often watched the minelaying or, if this was not the case, sooner or later they had a mine in their nets and were able to determine the presence of a field; reported to enemy minesweepers rendering the minefield mostly ineffective. For our submarines which had to crisscross the North Sea, these mine barriers were a major handicap. At the request of the Commander of Submarines, Admiral v. Pohl ended minelaying of this kind at the end of 1915.

May had a total sinking result of 107,515 tons - 53 vessels. The warship successes of "U21" in the Mediterranean are not counted here. In addition to the boats mentioned, "U27" with 10,141 tons, "U30" and "U41", which had worked on the west coast, were also involved. "U30", KptLt. von Rosenberg-Gruszczynski, sank 7 steamers of 21,177 gross register tons at the beginning of May. "U41", KptLt. Hansen, sunk 9 steamers (25,092 tons) on the west coast between May 25 and 29.

It was all the more satisfying to look back on the activity in May when there was no loss of submarines.

## June 1915

June should be less favorable in this regard. This month, the enemy came up with various countermeasures that had been prepared in the meantime, and not all of the boats were affected by them. This counteraction occurred particularly in the area of the North Sea and the English Channel, the west coast of Great Britain remained temporarily free of it for the time being.

On June 5th we lost "U14" in front of Aberdeen, which was sunk by the armed fishing trawler "Hawk", who was accompanied by 7 other fishing trawlers. The trick was when the submarine approached, the fishing trawler pretended to be harmless and unarmed and then suddenly fired when the submarine challenged the trawler crew to abandon ship from a distance of 1,500 meters. "U14" had only a 3.7 cm cannon and therefore could not reach the trawler from a great distance. When "Hawk" fired, "U14" could only attempt to escape the attack by diving. This would also have provided a way to save the boat if a newly installed vent had not accidentally closed in the boat's bow so that "U14's" forecastle remained above surface. This part of the boat was so badly injured by the trawler's gunfire and ramming that it had to surface again where the boat's destruction was completed. The commander, Oblt. Hammerle was killed. The commander of the I Submarine Half-Flotilla, KptLt. Mühlau, who had gone with him, was captured by the English. Before the loss occurred, "U14" had sunk 2 neutral steamers with contraband goods.

On June 23, "U40", KptLt. Fürbringer, was torpedoed and sunk by the English submarine "C24", which was submerged in tow by the armed trawler "Taranake". English procedure: When the German submarine approaches, the English trawler releases the tow to "C24" which is now able to maneuver submerged near the trawler. The latter acts harmless and stops upon receiving the German submarine's warning shot, immediately having its crew appear to abandon ship enticing the German submarine to unsuspectingly approach; this allows the enemy submarine the desired opportunity to shoot down the German one. Experienced submarine commanders sank such trawlers after their crews abandoned ship by artillery fire at greater distances and using zigzagging courses when firing; the commanders of "U14" and "U40" were on their first war cruise when they were so surprised by the enemy's cunning.

The two boats were soon avenged by "U19", KptLt. Kolbe and "U25", KptLt. Wünsche. The former sank 19 fishing trawlers in the North Sea in addition to 2 sailing ships and 4 cargo steamers; the latter 1 cargo ship, 8 fishing trawlers "U25" was rammed by a steamer (submarine trap) during an underwater attack, but got away with a bent periscope. At the time of its departure "U19" had provided reconnaissance services for the auxiliary cruiser "Meteor" and accompanied it up to latitude of Sognefjord, the latter tasked with a minelaying operation in the White Sea under the command of the KKpt. v. Knorr.

It is worth mentioning at that time boats which where were not yet ready to participate in the trade war were occasionally sent to stalk warships in the North Sea. In mid-June, four boats were laid out in front of the Firth of Forth in an east-west line. These efforts led to the torpedoing of the English armored cruiser "Roxburgh" by "U38" on June 20, 1915, under KptLt. Max Valentiner, who has occasionally been previously mentioned and later became well known through his actions in the Mediterranean. Unfortunately, the torpedo hit in the bow, so that "Roxburgh's" speed remained undiminished, the ship was able to avoid further attacks and was able to reach the port.[70] 3 cargo ships and 16 fishing trawlers, apart from the torpedoing of the "Roxburgh", would be Valentiner's tally on this cruise.

On June 14, the two small minelaying boats "UC1" and "UC2", destined for Flanders, came to the Ems while in transit to Flanders. They were led by Oblt. v. Werner and Oblt. Mey. Both boats had to stay in Emden for a few days because they had suffered various damages from bad weather at sea. At the end of the month they went on to Flanders. Oblt. v. Werner was successful there for many years. Oblt. Mey, also an excellent officer, would not return from his first cruise, which he made from Flanders to Lowestoft in late June 1915.

**"U30's" Mishap.**

June 22 was a particularly unfortunate day for the strike forces posted on the Ems.

At lunchtime, the mine clearing ship "Bielefeld", which was carrying out a mine control trip in front of the river mouth, was torpedoed by an enemy submarine (it could still be towed into the dock) and in the evening the - albeit only - serious accident occurred among submarines being supported at the Ems during preparations for

---

[70] HMS Roxburgh remained under repairs until April 1916.

their war cruises. "U30" flooded while undergoing a "pressure test" in a 40 meter deep channel in the vicinity of Borkum-Reede, due to a sight glass burst in the vent pipe of a diving tank allowing a three finger thick water jet penetrate into the boat. By a chain of unfortunate circumstances it was not possible to switch off the jet using valves that were provided for this purpose, or to bring the boat to the surface by blowing it out with compressed air before critical amounts of water had penetrated; only the stern of the boat came up steeply for a while. Unfortunately, an immediate attempt to tow the boat onto shallow water failed, only the commander, KptLt. v. Rosenberg-Gruszczynski, the helmsman and a sailor from the conning tower were rescued before the submarine entirely sank. The remaining part of the crew could not be rescued despite the greatest efforts of the rescue ship "Vulkan" which wase immediately brought in from the Baltic Sea. The final lifting of the boat was carried out by the Nordic Salvage Association by means of its salvage ships "Baltic Sea" and "North Sea". Due to the strong current at the scene of the accident and the short periods of still water, working from the great depth of water (40 meters) was very difficult for the divers. Mastermate Hünecke from the III Submarine Flotilla performed services that were not previously considered possible by the Nordic Salvage Association. On August 27th, the boat was brought up to Emden - very gradually, taking advantage of the tidal range. The brave men who died in the accident were buried with great solemnity.

"U22" and "U33" returned from successful long-distance cruises on June 23 and 24. "U33" under KptLt. Ganßer sank 2 steamers in the north-west of Scotland and brought in another steamship as a prize. There were also about a dozen steamers examined for contraband, but these had to be released due to the restrictive provisions placed by political and war authorities. It was clear these steamers would travel in the interests of the enemy on later voyages, so one can measure how much their sinking would have served the purpose of the submarine war. Of course, such examinations and releases of steamers were made in exceptional numbers on all submarine cruises. In the interest of narrative clarity, mention of all individual cases is omitted; the example is intended to demonstrate how enormous the effectiveness of submarine warfare was restricted by adhering to the strict provisions of the submarine blockade order published on February 4, 1915, and further removed from its intent.

"U22" had had the difficult task of picking up escaped prisoners of war submarine commanders from a certain point on the Irish coast. KptLt. Hoppe completely executed the task assigned to his boat unfortunately he could not take the officers with him because their escape from the camp had been discovered. During the operation "U22" sank 3 cargo steamers and a fishing trawler.

"U32", commanded by Freiherr v. Spiegel, brought a prize to List on June 23.

"U34" under KptLt. Rücker was on the way to the west coast from May 25 to June 13. The boat sank a Norwegian bark, the English steamer "Inkum" (4,747 tons), the French steamer "Penfeld" (764 tons) and 3 fishing trawlers.

The biggest result of the month of June was achieved by "U35" under KKpt. Kophamel. The boat traveled to the west coast of Great Britain from May 29 to June 23. 14 vessels with 23,530 gross register tons were his tally.

June registered a total of 114 ships with 115,291 tons, of which 14 ships with 14,080 tons were from Flanders' boats.

### Flanders

The small Flanders boats had been trained slowly and surely by their leader, KKpt. Bartenbach. So far we have seen them working in the Hoofden, avoiding the Dover-Calais passage, which had been considered blocked since the experience of "U32" (see page 50). In the first half of June, their activity there intensified: The minelayer "UC11", commander Oblt. Walter Schmidt, successfully carried one mine load after the other to the mouth of the Thames. (See Fig 9-

13.)[71] "UB6", Oblt. Häcker, sank the English steamer "Saidieh" (3,303 tons); "UB16", commander Oblt. Hans Valentiner, on June 3rd and 4th three English fishing trawlers north of Nord-Hinder Lightship; "UB10", Steinbrink, on June 7th a Belgian 4,000-ton steamer at Elbow-Buoy; "UB2", Oblt. Fürbringer, on June 9th six English fishing trawlers at Brown Bank; "UB16" on June 12th at Shipwash Lightship the English steamer "Leuctra" (3,027 tons)[72]; "UB13", commander Oblt. Becker, on June 17th an English guard steamer[73] at the Outer Gabbard Lightship, on June 19 the English steamer "Dulcie" (2,033 tons) at the Shipwash Lightship. It should be remembered that all these results were achieved in front of the largest and most important war and trade port in England, in the area of the strongest countermeasures and among the Thames shoals under navigational difficulties.

On June 21, KKpt. Bartenbach, commander of the Flanders Submarine Flotilla, decided to try again through the Calais narrows. "UB6" solved the question. The boat completed the passage at night without encountering any significant obstructions. As a result, on June 27th, the decision was made to dispatch four boats past the straits and set them on the Dover-Boulogne steamer route. However, this operation was unsuccessful; two boats did not get a shot after they had easily passed the narrows due to fog, bad weather and strong countermeasures, one boat encountered a new barrier between Colbart-Sand and Gris-Nez and returned, the fourth suffered a serious machinery accident near Ruytingen-Sand and after an 11 day absence was brought in with great difficulty. It must be remembered these boats had only a motorboat engine of 60 horsepower and that their surface speed was hardly superior to that of the strongest current in the Dover-Calais narrows. They could only get there traveling with the current, an unfavorable current

[71] All mining illustrations are drawn from Rene Jouan, "La Guerre Sous-Marine. Vol. II. & III." Payot, Paris 1934 and 1935. Translations of the German Official History "Handelskrieg Mit Ubooten" volumes referred to earlier in the text.

[72] Footnote in original. On June 23, the English Steamer, "Tunisiana" (4,220 tons) in front of Lowestoft.

[73] Postwar records confirm the attack was unsuccessful.

meant they had to await the appropriate time mostly on the seabed; for these boats, just passing the narrows was a masterpiece. (see Fig 10. for UC minelaying in the Narrows.)

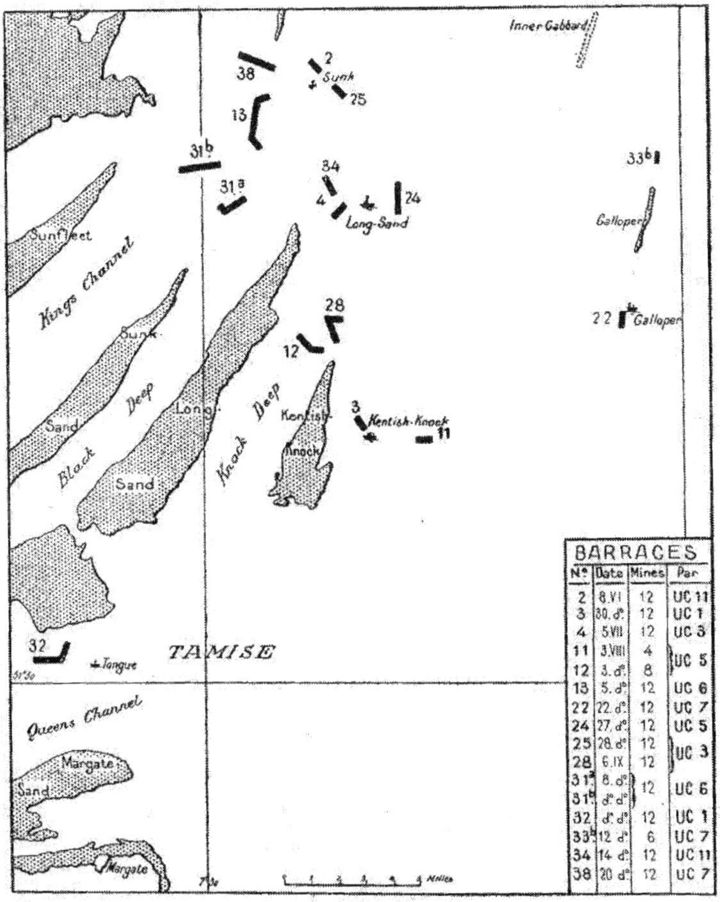


Fig 9. UC minelaying at the Thames estuary.

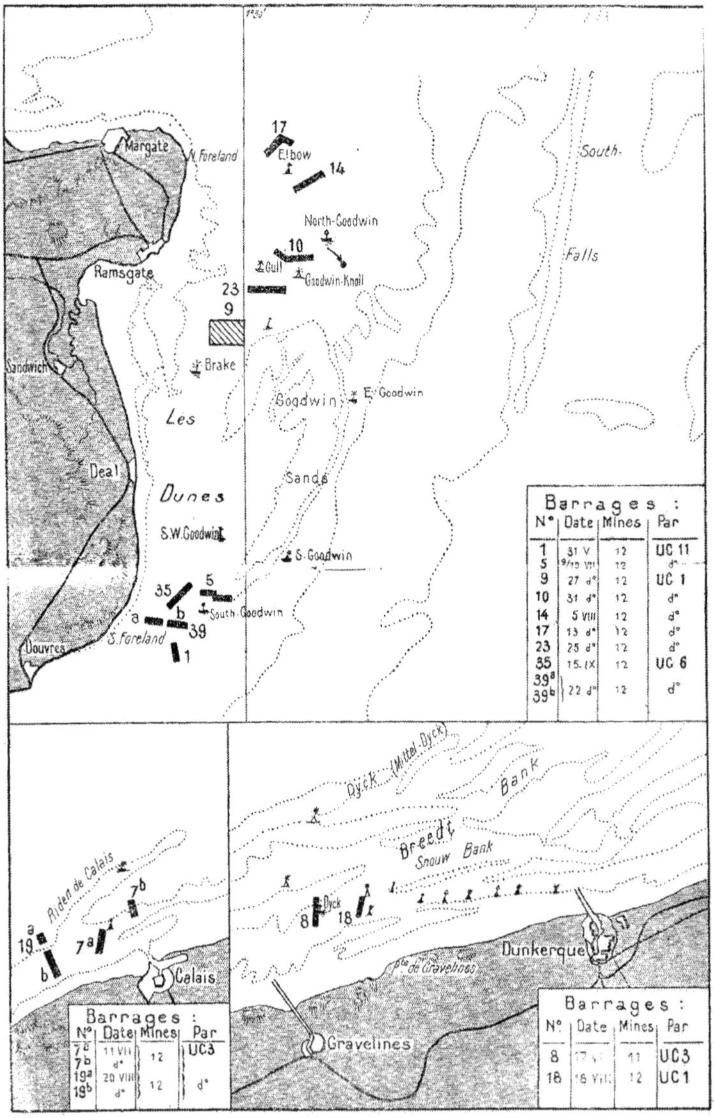


Fig 10. UC minelaying near Dover, Calais and Dunkirk.

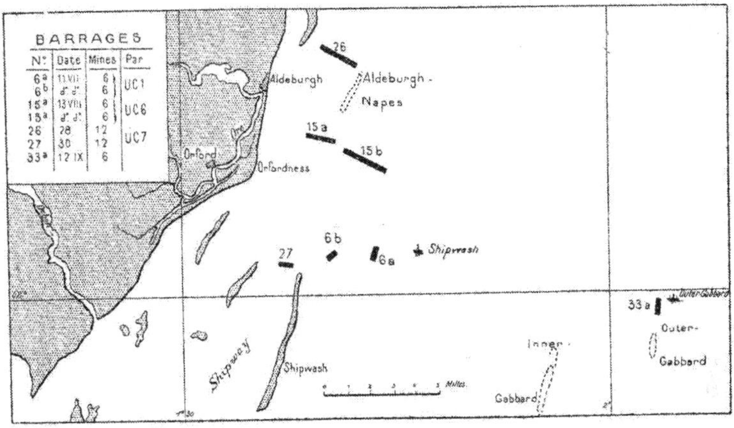


Fig 11. UC minelaying at Oxfordness.

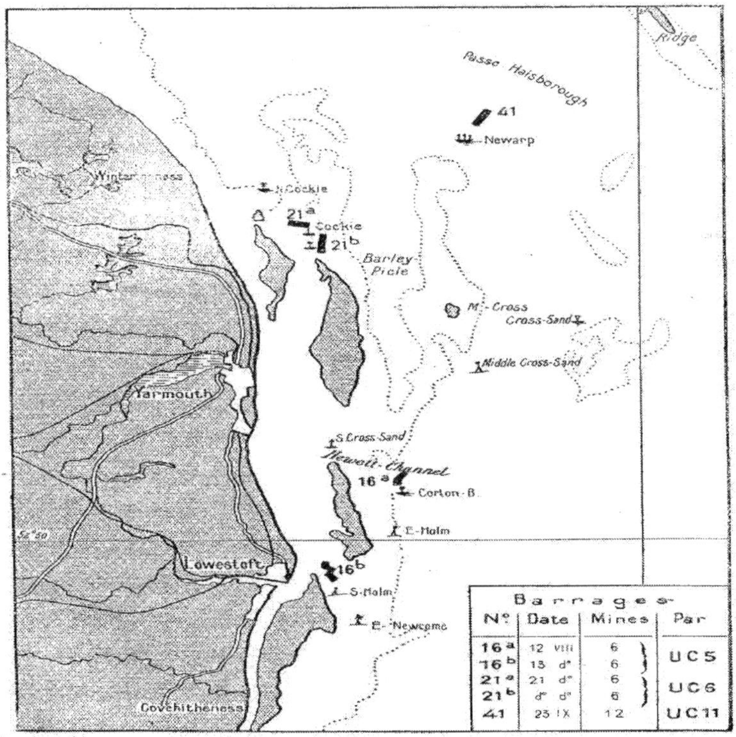


Fig 12. UC minelaying at Yarmouth and Lowestoft.

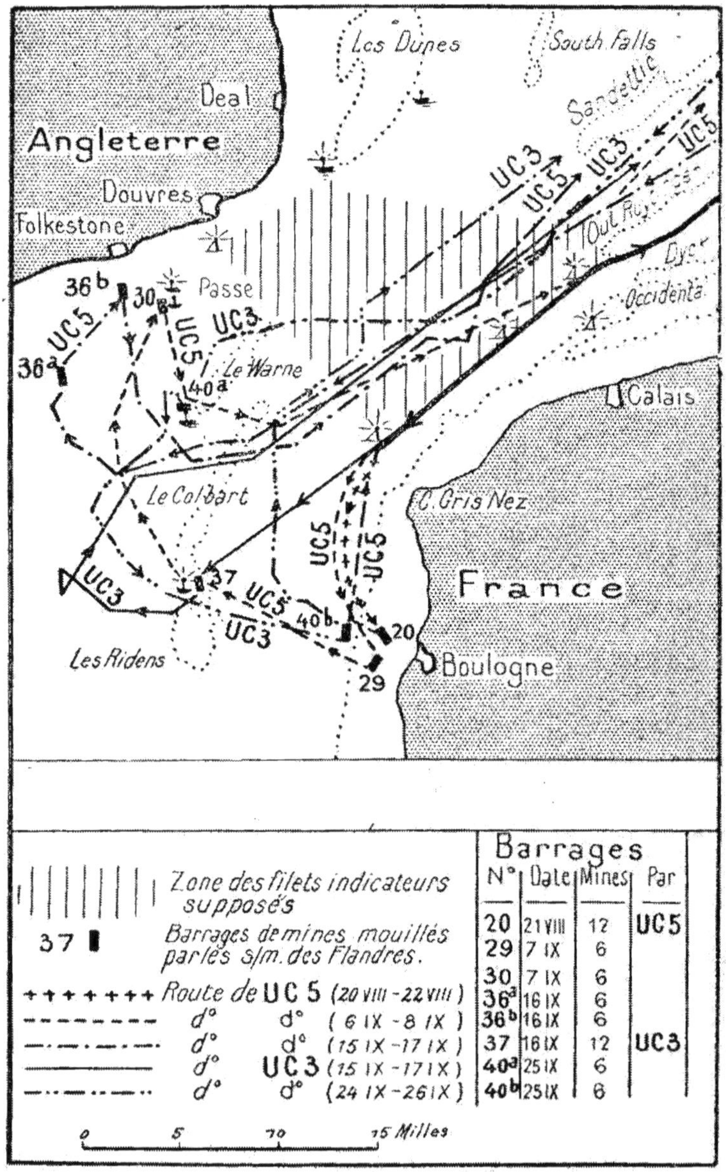


Fig 13. UC minelaying of Channel Narrows ports.

## Mediterranean Sea

In addition to Flanders the small submarines also made a name for themselves in the Adriatic in June. "UB15", which had passed to the Austro-Hungarian Navy, sank the Italian submarine "Medusa" on June 10th. Anticipating events somewhat, it should be mentioned here: On July 7, the Italian armored cruiser "Amalfi", accompanied by a destroyer division, was about to launch a bombardment against the Austrian coast, fell victim to another submarine attack.[74]

## July 1915

On July 4, "U9", KptLt. Spieß, and "U10", KptLt. Stuhr, received orders to be available to the Baltic Sea Commander and immediately commenced crossing to the Baltic Sea. Their exit from the war in the North Sea was not particularly significant given the extraordinary dangers these old petroleum boats were exposed to as enemy countermeasures increased.

On July 6, "U24" returned from a cruise to the west coast under KptLt. Schneider. 5 English cargo steamers, 3 English, 1 Italian, 1 Danish, 1 Norwegian sailing vessel was his tally. He had had some neutral steamers throw contraband cargo overboard. (28,171 tons sunk.)

"U25", Wünsche, whose engine range radius did not allow travel to the English west coast, was from June 28 to July 12 working in the North Sea. He sank a British schooner, the one Russian, English and Norwegian freighter and two British trawlers; he was in a surface battle with 3 armed English trawlers and severely damaged them. The cruiser "Hampshire", however, escaped his attack on July 1 in the Moray Firth due to a torpedo failure. Wünsche was fortunate to escape the furious submarine hunt (see Jellicoe, page 231, paragraph 4), which then deployed against his boat.

74 "UB14" was the attacker in this instance.

Significant sinking results on the west coast were achieved this month by KptLt. Schwieger ("U20") and KptLt. Forstmann ("U39"). Schwieger was at sea from July 3 to July 16; he sank a Russian four-masted bark, 2 English and 2 Russian steamers (9,462 tons). Forstmann made his way through the passage between Fair-Island and the Orkneys and headed for the Scilly-Isles via St. Kilda and Valencia. Between these islands and Ushant he sank 10 steamers and 4 sailing vessels of 35,955 tons in a few days. He was able to expend all his torpedoes and artillery ammunition, only regretting not having twice the amount of ammunition on board. From July 14th to 19th, "U41", Hansen, was enroute. The boat was supposed to conduct trade warfare on the west coast when it encountered 3 Pelorus-class cruisers north of the Shetlands. The torpedo shot against the middle cruiser failed due to the zigzag course of the ships; during the attack, the boat was rammed by one of the wing cruisers. Both periscopes were rendered unserviceable and "U41" had to turn return. A Russian steamer was sunk before and after the described incident; both were traveling between England and Arkhangelsk.

As in June, the enemy countermeasures on the west coast remained modest in July. It was a pity that many boats could not be used because of the unreliability of their machinery; these included the still existing petrol boats "U6", "U16", "U17, also" U23 "," U25 ". The two-stroke engines of these latter two boats still needed improvement. The fact that "U24", which had the same type of engine, had previously met all requirements, was due to the fortunately improved engine function on this boat thanks to the outstanding achievements of Marine-Engineer Peters, who had been the boat's chief engineer since commissioning. "U19" was not able to go to the west coast at this time for the same reason; this boat had the first experimental four-stroke diesel engines from M.A.N. (Maschinenfabrik Augsburg-Nürnberg). Only later was it possible to develop this boat's engines for longer range requirements through a fundamental conversion (installation of a modern control system). Some other boats, e.g. "U28" and "U32", temporarily dropped out of trips to the English west coast due to the state of their machinery. In July these circumstances were unfavorable for overall success; because working in the North Sea was made more and more difficult by enemy countermeasures. We know now from Admiral Jellicoe's book that at that time there was a systematic stalking of our submarines with patrols and traps in the North Sea.

"U23" under Oblt. Schulteß, who had already successfully carried out two North Sea cruises, fell victim to the same trick as "U40" (see page 60). This time it was the English submarine "C27" which was towed by the armed trawler "Prince Louise" (see Jellicoe, page 233). "U23" felt so safe that it stopped near the abandoned fish steamer seeing the hurry in which the English crew left their trawler after the warning shots. This gave the enemy submarine an opportunity to fire two torpedoes against "U23", the first failed, the second hit (July 23, 1915).

On July 26, "U16" under KptLt. Hillebrand almost fell victim to the same "Taranaki" trap steamer that sank "U40" with the help of submarine "C24". This time, however, the enemy submarine did not come into play, "Taranaki" opened fire, after "U16" fired a shot in front of its bow with its 5cm cannon at a distance of 1200 meters, immediately firing from two guns, at the same time replacing the originally flown Danish flag with the English war flag. "U16" was hit several times, but was still able to get the boat under water; various important diving mechanisms became un-calibrated, "U16" dove too deep that it had to emergency surface and rose again out of the water. Once there it received a new hit. Again, it went to a depth of 76 m, the boat bottoming aground. It was only after an extensive effort that Hillebrand was able to detach the emergency ballast and move the heavily flooded vessel off the bottom, proceed underwater and thus run out of the enemy's sight. The mandatory surfacing turned out to be very difficult, since the diving tanks and their ballast air lines were mostly shot through. Only by simultaneously blowing out the still submerged diving tanks and part of the petroleum bunker was it possible to make the boat buoyant again. With a list of 8° to 10° the trip to Helgoland started, where it was a happy day when they arrived.

If, in this case, the trawler “Taranaki” was wrong when he claimed the submarine had been sunk (Jellicoe wrote the submarine was sunk.), another fight took place on 27 July between the armed trawler "No. 830 "and" U41 "under KptLt. Hansen near St. Kilda on the west coast and we also got off lightly. "U41" received a hit in the conning tower, which forced the boat to cut its operation short, which KptLt. Hansen, one of the great daredevils among our submarine commanders, did very reluctantly. Nevertheless, Hansen achieved the following sinking result on this venture: 5 English fishing trawlers, an English steamer with wheat from Arkhangelsk to le Havre, an American steamer with flax from Arkhangelsk to Belfast and a Norwegian steamer which had been captured by the English (10,776 tons).

"U6", KptLt. Lepsius, brought two prizes to List on July 24 after examining 41 vessels (including 21 steamers) between Skagerrak and Firth of Forth; only 9 sailing ships and one steamboat could be sunk under the applicable regulations.

On July 24, "U36", KptLt. Graeff, was sunk by the armed collier "Prince Charles", which cruised as a submarine trap. The same procedure here: To entice the submarine closer the steamer crew willingly abandons ship at the first warning shots. As the submarine approaches, a previously hidden gun opens fire. "U36" tried to escape this fire by diving, but was no longer able due to hits already received. Before this misfortune occurred, Graeff sank the following vessels: 9 English fishing trawlers, a Norwegian sailing ship with wood pit props, Norwegian steamer "Fimreite" with 6,500 tons of iron ore from Canada to England, the French steamer "Danae" and Russian steamer "Rubonia", also the American fully rigged sailing ship "Pass of Balmaha" – carrying cotton on the way to Arkhangelsk - sent through under Helmsman Mate Lamm as a prize to Cuxhaven (total result 12,214 tons). The "Pass of Balmaha" later became the "Seeadler", which, as a German auxiliary cruiser under the leadership of Count Luckner, was to acquire everlasting laurels.

On July 28, the Imperial Navy Office made available, at the Ems, the steamer “Rugia”, from the Hamburg-America Line, to the III and IV Submarine Flotilla as an accomdation ship. This ship became the focal point for important information exchanges of war experiences for the numerous Ems submarines. Later it became very important for all German submarines because its radio station became the control center for all submarines in the North Sea. The station also kept in touch with Mediterranean boats. At night important operational news went to the boats from here, at other time schedules it received reports from boats in the Atlantic, Mediterranean and North Sea; commanders at home feverishly awaiting patrol updates.

July brought a total result of 86 vessels with 98,331 tons. The Mediterranean contributed one vessel, 5,601 tons. This was the French transport steamer "Carthage", which fell victim to a torpedo from Hersing in front of Seddul Bahr on July 4th. "U21" traveled from Constantinople to the Aegean Sea for this venture.

In Flanders, the local submarine commander initially decided not to undertake further activities beyond the Dover-Calais narrows based on the experience he had gained at the end of June. In contrast, boats "UC1", "UC2", "UC3" and "UC11" experienced a lively, successful mine-laying activity in front of the Thames estuary.(See Fig 9.) In addition to its own trade war activity - 24 vessels of about 7,000 tons were sunk - the B-boats performed important services by prior reconnaissance of the minelaying areas. Here again Oblt. Steinbrink ("UB10") excelled by entering the Downs from the north (fairway between Goodwin sand and the English coast) and advancing almost to the Gull buoy, noting the Downs were closely guarded day and night by fishing trawlers and destroyers. "UC1", Oblt. v. Werner then performed several bold minelaying trips against traffic in the Downs.(see Fig 11.) "UC2", Oblt. Mey, which had departed on June 29 to lay a minefield off Lowestoft, was declared lost in mid-July.

## August 1915

This month can be described as one of the most eventful of the submarine war.

On August 3, "U22" under Lieutenant Hoppe left Borkum-Reede to head for Norway's Vestfjord (at 68° latitude). We had received news that an English auxiliary cruiser was permanently stationed there, northwest of Bodo, which had the task of seizing and sinking German ore steamers from Narvik. "U22" reached the fjord without incident and the boat had hardly taken its waiting position when the English auxiliary cruiser "India" (7,940 gross register tons and 340 crew) entered Vestfjord. A torpedo fired at 1600 meters hit the ship, which soon sank with 160 men. It was one of the rare cases where a submarine actually found the target it was targeting with the precision reported by an agent. Furthermore, "U22" sank the English steamer "Grodno"[75] on this trip north of Lofoten.

On August 4, KptLt. Wegener left for the West Coast with "U27". As early as August 10, Wegener again had a warship success (his fourth since the beginning of the war)[76] by inflicting a torpedo hit on an older English cruiser, probably of the Amethyst class, west of the Hebrides that day.[77] Wegener himself was unable to bring the news of this success home; we received it from the commander of "U38", KptLt. Max Valentiner, who later met with Wegener in the southern Irish Sea. Wegener along with his good crew fell victim to an atrocity by the English auxiliary ship "Baralong" on August 19 when the American steamer "Nicosian" was being sunk. "U27" was sunk when surprised by "Baralong" suddenly appearing behind "Nicosian" and sunk by artillery fire. The English flag went up on "Baralong" which had previously carried the American flag and side badges before it opened fire. Eleven men, including the commander, were able to get out of the submarine in time and swim towards the "Nicosian". Partly in the water, partly on board the "Nicosian", they

[75] Footnote in original. 1,955tons.

[76] HM "E3", HMS "Hermes", HMS "Bayano".

[77] The torpedo attack was unsuccessful.

were murdered in an evil manner on the orders of the captain of the "Baralong". With Wegener we again lost one of our most excellent submarine commanders. He was also popular and respected as a person; he had often delighted the circle of known submarine comrades in Borkum-Reede, when thick fog precluded any cruising activity, by sometimes serious, sometimes cheerful stories. - The exchange of notes which followed the atrocity is so well known that it need not be repeated here. It should only be mentioned the English government did not let the "Baralong's" captain, despite his crimes, answer to charges, even if other news said that he had been relieved of duties.

KptLt. Max Valentiner, who, as mentioned, met with Wegener for a joint venture in the Irish Sea, was particularly fortunate on this cruise. In five days he sank in the St. George Channel, and in front of the Bristol Channel and Ushant, 22 cargo steamers, 5 fishing trawlers and 3 sailing ships with a volume of around 70,000 tons. This huge result, the largest ever in the war in Britain's waters, was achieved primarily by using the 8.8 cm cannon in poor visibility; the torpedoes were only used for parting shots when a steamer had to be sunk faster due to a patrol's approach. You can see the importance of artillery was still considerable at this time due to the lack of counter-armament.

On August 4, "U34", KptLt. Rücker and "U35", KKpt. Kophamel, left Helgoland for the Mediterranean. They were only allowed to conduct trade war up to the entrance of the Channel, and then they should proceed unnoticed. "U35" sank three vessels up to that point with about 5000 gross register tons. On August 23, both boats arrived safely in Cattaro.

On August 5, "U24" and "U25" left Borkum-Reede. "U25", as mentioned, could not be dispatched to the English west coast due to the condition of its machinery, achieved the following result under KptLt. Wünsche off the Norwegian coast, on the Archangel-England steamer route and off the Skagerrak: one Swedish, four Norwegian steamers and a Norwegian barque sunk with contraband, 11 steamers examined had to be released again in accordance with the restrictive regulations; three Norwegian steamers had part of their cargo that was declared contraband was thrown overboard. On the return trip, both of "U25's" engines failed to such an extent that it had to be sought out by two destroyers north of Horns Reef and towed into the German Bight. The chief engineer of the boat, Naval Engineer Ladewig, had gotten all the performance that was humanly possible out of the weak machinery, and the boat had covered distances at the beginning of the war which would not have been considered possible. The boat found itself at the end of its combat career; it could only be restored for Submarine School use, where there is little need for long radius performance. KptLt. Wünsche took over the new boat "U70" - one of the submarines originally built for the Austrian Navy at the Germania shipyard. He returned to the front in early 1916 and led it from victory to victory to ever more brilliant results.

KptLt. Schneider with his old "U24", the engines of which, as mentioned, were better assembled, sailed north around England to the Irish Sea, where the boat was tasked with the destruction of a benzene factory near Harrington (in the northwestern part of the Irish Sea) On August 16, Schneider appeared early in front of the Firth of Solway and had finished his work of destruction quickly before anyone could fight back. Schneider then left the Irish Sea through its southern entrance, the St. George Channel, where he was attacked without being damaged on August 18 by a large English steamer with artillery,. Again, like the sinking of the "Lusitania", the area south of Old Head of Kinsale on the south coast of Ireland was to become the scene of an event that entangled us in new note exchanges with the United States.

Schneider had stopped the steamer "Dunsley" on August 19 48 nm to the west of Old Head Kinsale and was about to sink it after its crew left when a large steamer without a flag and neutrality badge quickly approached in the afternoon. Based on what Schneider had experienced on August 18, he sensed offensive intentions with this steamer, dived, attacked and fired a torpedo, from which the badly hit steamer quickly sank. "U24" was unable to establish the name of the steamer since the arrival of patrols was expected soon.

Only after "U24's" return was it determined to be the White Star Line's "Arabic"[78], which on the morning had left Liverpool for New York with 170 passengers and 250 crew. While the steamer had brought significant quantities of war material on its last trip from America, it now carried 8 million marks in English securities. This loss alone, in addition to that of a valuable steamer, was sufficient to trigger a similar storm of indignation in England as after the "Lusitania" sinking. In addition, several US citizens were on board, three of whom were said to have drowned. So England found the most willing ear for her outrage across the Atlantic.[79]

Count Bernstorff, without being officially authorized to do so, declared in America that "the submarine commander would be punished for exceeding his instructions." The "Arabic" case sparked fierce struggles at Grand Headquarters. Von Tirpitz and the Chief of the Naval Staff, Admiral Bachmann, with utmost energy took the position that the submarine war should not be abandoned due to America, while the Chancellor von Bethmann-Hollweg, who was seconded by Chief of the Naval Cabinet, Admiral von Müller, was in favor of assuring America the "Lusitania" case should be submitted to an arbitral tribunal and that submarine commanders had specific orders not to sink passenger steamers without warning and rescue passengers. Although the Emperor had decided on August 26th to agree with Tirpitz and Bachmann, the Chancellor brought about a reconsideration of the Imperial decision towards his position on the 27th. The fateful dispatch to America went off. Bachmann, who had been ignored, resigned his position; retired Admiral v. Holtzendorff became his successor. Grand Admiral v. Tirpitz relationship with those higher up continued to deteriorate. It was clearly felt within the submarine arm that the submarine war in the North Sea - even in its currently very limited form - was slowly but surely being

---

[78] Footnote in original. 15,801 tons.

[79] Footnote in original. Schneider's overall result on this cruise was: 6 steamers, 1 sailing ship sunk, 1 steamer seized off Lister Deep. (27,743 tons.)

buried, especially since on August 16 boats "U33" and "U39" set out for the Mediterranean. In addition to creating favorable operational results against hostile trade in the Mediterranean, this posting also promised to avoid further involvement with America.

By the way, a command to finally put an end to the trade war in the North Sea did not yet reach the submarines, so the sinking results in September were not influenced. On August 30, an order was issued not to sink any small passenger steamers without warning and rescue crews until further notice. This command was simply not executable. How should a submarine commander distinguish a "small passenger steamer" from a cargo steamer through his periscope? Fleet Commander v. Pohl protested the order on behalf of his submarine commanders by stating this position. The Emperor did not respond to this and rejected the Fleet Commander's position.

One day in the month of August, on the 8th, enemy forces appeared in the German Bight; they came partly from the "Grand Fleet" from Scapa Flow, partly from Harwich and with the purpose of intercepting the auxiliary cruiser "Meteor" on its way back to the German Bight. Under the command of the KKpt. von Knorr, "Meteor" had heavily mined the Firth of Moray and sunk the English auxiliary cruiser "The Ramsey"[80] ("Meteor" had been escorted by "U17", KptLt. Walther when it departed). The Harwich Forces were sighted on August 8th in the morning by our aerial reconnaissance northwest of the mouth of the Ems, whereupon "U32" was immediately launched against them from Borkum-Reede. Likewise, "U28", which returned from a successful long-distance cruise - it had sunk ten English steamers (22,235 tons) - was informed of the location and course of the enemy. This boat was not used against the enemy armed forces, but sighted a neutral fishing vessel with KKpt. v. Knorr and "Meteor's" crew onboard and brought them to Wilhelmshaven. Knorr had to scuttle "Meteor" itself when he sighted the English cruisers, who passed in close

[80] Footnote in original. 1,443 tons.

proximity to the fishing vessel, and they also became aware that the survivors of "The Ramsey" were on board her. The English Commodore, however, did not risk stopping to take on these men because he had a German airship in sight and at the same time thought he was being attacked by a submarine. The "Meteor's" crew was not captured due to these events. "U32" was unable to find the Harwich Forces and was recalled back on the 10th to the Ems.

On August 16, "U17", KptLt. Walther, brought the Swedish steamer "Göteland" as a prize to List after he had sunk four steamers, two fishing trawlers and a sailing ship.

On August 27, KptLt. Forstmann left Helgoland for the Mediterranean with "U39". The boat's trip to Cattaro was quick in the good weather; it arrived on September 15th. Forstmann sank three cargo steamers between Cartagena and Algiers - a total of 6,616 tons. Forstmann was eventually to achieve a sinking result of over 400,000 tons in the Mediterranean.

On August 28, "U33" under KptLt. Ganßer set off on a cruise from Borkum-Reede to the Mediterranean. On the way there, Ganßer sank four English steamers and two Norwegian sailing ships (17,359 tons) to Cap Finisterre in Spain, arriving on September 16 in Cattaro.

As far as it is known, the sinking result in August was:

101 ships of 174,523 tons, of which Flanders contributed 30 vessels with 89,35 tons[81], 1 steamer of 11,117 tons in the Mediterranean. This last was the large English transport "Royal Edward"[82], a victim of "UB14" under Oblt. v. Heimburg near the island of Kos,. At that time Heimburg was crossing from Cattaro to Constantinople, during which the boat was relying on Orak Bay in Asia Minor to extend its operation in the Aegean (see footnote page 53, for information on "UC15").

[81] Foonote in original. Of which 7675t from "UB10", Oblt. Steinbrink.

[82] Footnote in original. 11,117tons.

In Flanders, minelaying activities were directed against the trade routes between the English coast and surrounding shoals ("UC1", "UC3", "UC5", "UC6" and "UC7" were involved). After a long time the Flanders Submarine Flotilla booked the first real success beyond the Dover-Calais line, "UC5" was the first boat to pass the Dover-Calais narrows and laid mines off Boulogne.(see Fig .13) This was achieved by Oblt. Pustkuchen. It should be noted that in August the five C-boats mentioned made 16 minelaying trips sinking 8,400 tons of English vessels alone. The commanders of the C-boats not yet mentioned were: "UC3", KptLt. Weisbach; "UC6", Oblt. Graf v. Schmettow; "UC7", Oblt. Wäger. From mid-August onwards, the Flemish coast was continually disturbed by enemy surface forces. The B-boats were therefore heavily stressed by use in outpost positions off the coast; the narrow passages between the shoals off the French-Belgian coast were blocked by C-boats with mines against enemy monitors. August 23 saw enemy monitors bombard Zeebrugge; however the available boats were not able to attack these ships.

"UB4", which had left Yarmouth on Friday, August 13[83], was declared lost at the end of the month. The commander was Oblt. Groß, who earned a reputation as an excellent officer in the first months of the war as Adjutant to the III Submarine Half-Flotilla.

## September 1915

Due to sending four large 30s boats to the Mediterranean Sea and the partial withdrawal of petroleum boats ("U9", "U10") from the North Sea, it became very quiet in the submarine bases there. Above all, this related to Helgoland, from where operated only "U6" under KptLt. Leptius, "U16" under KptLt. Hillebrand and "U41" under KptLt. Hansen who was lost in September. "UC9" was temporarily based at Helgoland from September 1 to 21; on September 21 it went to Flanders. Most of the submarines were still at the Ems,

[83] Footnote in the original. Friday as a departure day was always a bad omen for submarine commanders; a Friday that fell on the 13th was particularly unpopular as a sailing day.

where the following boats were based in September: "U20", "U22", "U24", "U28", "U32", "U43", "U44".

These last two boats had entered the front at the end of August under the command of KptLts. Jürst and Wagenführ. There were boats of a new type from the Gdańsk Shipyard (still under construction in peacetime). They were characterized by excellent underwater properties, but disadvantaged in having only 1000-horsepower diesel engines, which were too weak for the boat size so that practically no more than 13.5 knot top speed was reached and the engines often broke under overload. Given their size, the boats could easily have been equipped with 1200 horsepower engines. It was very painful for the brave commanders who led these and succeeding boats up to "U50" due to the lower engine power of their boats, they often had to lag behind the performance of their comrades who commanded faster and more reliable boats.

"U20" under KptLt. Schwieger carried out a successful cruise (36,526 tons) to the English west coast again this month; he advanced south to the Gironde estuary on the French coast. Nine steamers, six English, two French, one Russian, were his victims; one of the English steamers was an auxiliary cruiser of 10,000 tons.[84] "U20" achieved a mutual wireless reception range with "Arkona" of 575 nm on this trip, thereby setting a new record.

KptLt. Schwieger had sunk 115,000 gross register tons in five long-distance cruises, one to Le Havre, four to Irish waters, and had only traveled 17,800 nm on these long-distance trips. Marine Engineer Szitnick the senior engineer of "U20", is commended for the machinery's performance; he has been on board since the boat was put into service and had always kept the machinery at an excellent level.

Likewise to the west coast went on September 14th "U41" under KptLt. Hansen. Referencing the English Merchant Shipping Losses

[84] The vessel sunk was actually the civilian passenger ship "Hesperian". Gernan records indicate the vessel was registered as an auxlilary cruiser. The misidentification resulted in the ship being torpedoed without warning.

- published in August 1919 - on September 23 and 24, Hansen sank 4 English steamers displacing 19,692 tons south of Fastnet Rock. Then on the 24th the boat succumbed to an attack of the submarine trap "Wyandra" (aka “Baralong”). Here again the crew skillfully abandoned ship and faked surrender to the submarine which approached believing the situation was safe and then sunk by sudden rapid fire which smothered it. The loss of Hansen, since the war began one of the most skillful and daring submarine commanders, weighed very heavily for us.

On September 9, "U6" under Oblt. Lepsius was dispatched to the Norwegian coast for trade warfare. Lepsius sank three sailing ships with pit wood to England and sent the Danish steamer "Randulf Hansen" as a prize to the German Bight. Unfortunately, this boat should also come to an abrupt end on September 15th. We read about this in Jellicoe's book: "September 12: Submarine "E16", which was part of the Grand Fleet submarine flotilla, was sent from Aberdeen to the Norwegian coast to combat an enemy submarine that was believed to be operating in these waters. September 15 "E16" sighted and sank an enemy submarine off the Norwegian coast with a torpedo. This success, one of the first to be achieved by a submarine operating against its peers, was very encouraging after the somewhat similar success of "C27" in July." — It should be noted here that the activities of "U6" had become known to the English Fleet command through neutral steamers, who had been stopped and investigated by "U6". Nevertheless, it was a special stroke of luck for the enemy that he was able to find "U6", which at the time of the sinking was not working on a merchant vessel, but proceeding alone in the open sea - at the same time charging the battery accumulator. This was probably only possible through the widely visible white smoke that these boat engines emitted and which made their situation virtually impossible in areas with a strongly developed enemy defensive system. The intention to pull the last petrol boats "U16" and "U17" still stationed in the North Sea out of this war zone as well as "U9" and "U10" could therefore only be approved. - Lepsius was one of our oldest submarine officers. In 1910 he been a watch officer on "U6" and was therefore closely linked to the boat and crew - two officers and three men from "U6" were recovered by the English submarine.[85]

More fortunate than "U6" was "U16" under KptLt. Hillebrand, departing on September 19, operated again in the treacherous North Sea before this boat also finally moved to the Baltic Sea. Hillebrand sank a steamer, three sailing vessels and a lighter and brought in a steamer as a prize. After he handed over "U16" to the Submarine School in October, Hillebrand took over the new boat "U46" at the end of 1915, with which he had many successes against the enemy until the end of the war.

When a mining operation of our small cruisers of the II Reconnaissance Group took place in the northern Hoofden from September 11th to 12th, five Ems submarines, "U19", "28", "32", "43" and "44", were deployed in positions in which they could support our surface forces. Since the mines were laid quite a distance from the enemy coast, our ships did not come into contact with the enemy.

In Flanders, enemy forces repeatedly appeared off the coast in early September. The local submarine flotilla commander decided to maintain two guard submarines off the coast. Their positions were at the Middelkerke and Thornton Bank buoys. This use resulted in heavy wear on the submarines; damage to enemy shipping from then on had to be done mainly by C-boats. In mid-September, "UC3" and "UC5" went through the Calais narrows and laid mines near Boulogne. ;(Fig 13.) At the end of September, aircraft attacks against the incoming and outgoing boats began to take place in Flanders; "UB6" and "UC1" suffered damage as a result. As we see the minelaying activity of C-boats in Flanders develop successfully in August and September it is interesting to note the submarine establishment and technology had been trying to create minelaying submarines since 1910. These efforts were unsuccessful at the time. Upper level decision making only had in mind the use of the

[85] Lepsius was not among the survivors.

submarines together with the High Seas Fleet, mining by submarines was not considered very promising.

In the Mediterranean, where the focus shifted towards submarine trade warfare, a number of successes were achieved in September.

At the beginning of September, as mentioned, Forstmann sank three steamers totaling around 10,000 tons when he entered the Mediterranean between Cartagena and Algiers.

"U34", cruising the southern Aegean Sea from September 1 to 17, 1915, sank the English steamer "Natal Transport" 4,107 tons and the French auxiliary gunboat "Indien" 800 gross tons on September 4.

"U35", sailing from August 31 to September 22 in the Aegean Sea, sank three ships with 10,577 gross tons.

At the end of the month (from September 25th) "U39" was already successfully operating in the Eastern Mediterranean. In September the boat destroyed two English steamers with about 8,000 tons and an Italian steamer of 2,000 tons.

"UB14", v. Heimburg, torpedoed the troop transport “Southland” (11,899 tons) in the Aegean Sea on the way from Orak Bay to Constantinople on September 2; the steamer escaped its fate at this time, which ended on June 4, 1917 140 nm northwest of Tory Island through another German submarine.

"U21", Hersing, was active from August 29 to September 26 on operations outside the Dardanelles and towards Salonika. Due to various adverse circumstances, he was unable to achieve success. A re-entry by the Dardanelles was tantamount to risking the boat due to new minefields blocking the passage. Hersing requested and was granted permission to return to Cattaro by Vice Admiral Souchon, Commander of the Mediterranean Division in Constantinople. He arrived there on September 22nd.

Successes were also recorded in the Black Sea.

"UB8", Oblt. Voigt, sank four Russian sailing ships from a cruise to the Crimean coast, on September 11 he shelled a warehouse on Lake Uzunlar, later the Chauda lighthouse.

"UB7", Oblt. Werner, sank the English 6,000 ton steamer "Patagonia" enroute to operations near Odessa.

No boats to conduct trade warfare were dispatched to the North Sea after September 20. The Kaiser issued a command on September 20, 1915, to the Commander of North Sea Submarines and the Flanders Submarine Flotilla to end any type of submarine warfare on the west coast of Britain and in the Channel. Concluding the struggles described in the "Month of August", the intentions of the Reich Chancellor and Chief of the Naval Cabinet, which aimed at a definitive understanding with America, had now completely prevailed at headquarters.

This decision was less heavily felt in the North Sea, because with only 8 boats currently available there ("U38" was already assigned to the Mediterranean), vigorous trade warfare - especially since bad weather set in during the winter months - was not possible there as it was for Flanders, where the number of boats had risen to 17 and the long winter nights were more favorable than unfavorable in an operating area that was more protected against heavy weather. A consolation for Flanders was partly that at least the mine boats could carry out their activities unhindered, since this did not fall under the term "trade warfare". The mine actually rose in value as a result of the trade war cessation; we shall see later the inactivity imposed on North Sea submarines even led to the construction of special mines for the large torpedo submarines that could be laid from the boat torpedo tubes. (England also used such mines on submarines.)

In the Mediterranean, the submarine trade war continued and increased in success until it came to take its revenge that a generous expansion of repair bases had not started as early as 1915; viz. until the boats, which were materially strained due to the crossing and continuous operations, had to be undergo lengthy repairs. For the time being, this was not yet noticeable; it was not until March 1916 that there was a noticeable influence on activity.

In September 1915, a total of about 52 vessels of 130,178 tons were sunk; about 50,000 tons contributed by the Mediterranean.

The second phase of the submarine war, "the 1915 submarine blockade" had come to an end.

## Conclusion

As bitter as it was, at the Front one had to realize in mid-September 1915 that the "submarine blockade" announced on February 4, 1915 had collapsed. In volume 1 we stated, in Grand Admiral v. Tirpitz 's judgment as well as part of the Submarine Command, the blockade commenced prematurely. Even if "U21" and the 5 other large boats had not been sent to the Mediterranean, the 13 large boats then available in the North Sea would not have been sufficient for the blockade to continue effectively. 4 of the 13 boats would not have been suitable for use on the English west coast without extensive alterations. As mentioned the boats temporarily available in Flanders could only be used in a limited sea area. We had bled to death regarding boats in the North Sea; stock had to be refreshed with new, powerful boats before operations continued.

If the "blockade" had started later, around July 1915, the extent of the current situation would not have occurred. 4 North Sea boats were lost on the first cruise of their commanders who had not experienced combat operations. They could have been gradually introduced into operations and not need immediately experience difficult situations, thereby decreasing the possibility of loss. In such a case, the petrol boats could have been assigned the easier tasks, which would also have made their loss rate (4 out of 7) smaller. It would then have been possible to methodically assign every commander to the most ideal use of the boats, thereby further reducing the overall loss. Approximately a dozen Flanders boats would have been ready at this later date to increase the initial impact.

However, even this more favorable overall disposition would not have guaranteed success if the political leadership folded every time before America, as happened after the "Lusitania" and "Arabic" case. The extent to which boat commanders were afflicted with all sorts of restrictive orders and instructions can be judged only by one, who by virtue of his official position, was forced for several months to support them and encourage their morale and continued exceptional performance in spite of our politicians who invented all sorts of restrictions designed to confer immunity on a number of vessels. It was clear most commanders experienced a serious conflict of conscience, not to mention the resulting considerable reduction in tonnage sunk. We used various examples to show how numerous steamers had to be released by submarines after being examined for contraband. Who was to say these steamers would not carry contraband on the next voyage? Most of the time, the investigation could only be carried out on the basis of the ship's papers; have they always been issued correctly? Could not contraband be hidden under the load? - The number of neutral steamer not sunk as a result of standard operating procedures may have been 300 in the period described. If the steamer is averaged at only 2500 gross register tons, this meant a total failure of 750,000 gross register tons, more than destroyed since the commencement of the submarine blockade. This simple calculation demonstrates the only way to guarantee success was through an unrestricted submarine war, where every enemy vessel is sunk and also all neutral vessels that came from or steamed to an enemy port, viz. were found in sea areas where these operations occurred. Later when the matter came down to unrestricted submarine war it came to a battle against global tonnage – the latter was almost entirely at the disposal of our opponents.

The numerous commands restricting the submarine war as it was originally intended are only partially listed in our description; in fact there had been so many over time that it was no longer possible for the submarine commander to remember them all.[86] It was not uncommon for the helmsman or other officer to come into the tower or onto the submarine bridge before the attack with a large file, where it was only through hastily "leafing" through these files was one able to determine whether the vessel concerned could truly be sunk. And often it happened, as the saying goes: "However well you did it, it was wrong". It was a painful, almost impossible task to make it clear to a very exhausted, successful commander returning from a long-distance cruise that he still had to answer for one or the other sinking knowing the effect this had to have on the commander in question. This duty fell heaviest on the Commander of North Sea Submarines, Fregattenkapitän.[87] Bauer, and his Chief of Staff KKpt. Lützow, which took these officers often away from their operational work. It must be said that with such a muddle created by all these orders and instructions, the final order to end trade warfare in the English Channel and English West Coast was for some a great relief, even for undoubtedly brave commanders.

Every campaign which is well developed in operational terms and successfully carried out, but devoid of clear political objectives, is still only a Sisyphean labor.

Let us briefly touch on the activity of our surface forces in the period under discussion.

Apart from the above-mentioned operations serving the purpose of laying mines in the free North Sea, the High Seas Fleet was cautious. The advantages and disadvantages of the mining endeavors have already been evaluated in the narrative. Discussing the reasons for this reluctance would diverge too far from our topic. In any case, it was very regrettable the "submarine blockade" could not be supported by more intense activity from the High Seas Fleet.

---

[86] Rear Admiral Andreas Michelsen indicates 146 specific submarine orders were issued during the war. For example No. 24 of June 5, 1915 prohibits the sinking of passenger ships. See his Der U-Bootskrieg, 1914-1918. Leipzig: von K.F. Koehler, 1925

[87] Commander in USN parlance. Hereafter FKpt. Hermann Bauer referred to earlier had been promoted.

Admiral v. Pohl experienced pressure to undertake fleet operations when the blockade's progress did not appear to have any lasting effect any time soon due to the High Command's restrictive orders and the loss of submarines.

Apart from the torpedo boats, support for submarine warfare by the actual High Seas Fleet was out of the question at the time. The Ems and Helgoland based submarines each had a half-flotilla of torpedo boats available since April. Their work was of great value to the submarines by helping them during egress/ingress.

The Ems Coastal Protection Flotilla under the command of the FKpt. v. Lessel and later FKpt. Ewers supported the local submarine flotilla in the most accommodating manner. The ships ("Hagen", "Beowulf", "Heimdall") gave the submarines the opportunity to improve their torpedo shooting by functioning as target ships off Borkum-Reede, the mine clearance and smaller vessels worked tirelessly to clear the entry and exit routes from mines. The mine barrage group under the command of KptLt. der Seewehr[88] Simonsen merits particular mention. They carried out many operations without any support from larger forces, often at great risk.

In order to fully understand the situation at the time and the further prospects for the use of the submarine, it is now necessary to briefly clarify the material status of the submarine at the end of the second phase of the submarine warfare. What were the operational submarines at the front? What could we expect to have in hand? a) Already in service, b) Boats completed or under construction.

In late September 1915, at the front there was:

> a) In the North Sea ten boats – “U19”, “U20”, “U22”, “U24”, “U28”, “U32”, “U38”, “U43”, “UC10”.

[88] Seewehr designates an older officer called from retirement.

b) Flanders fifteen boats – "UB2", "UB5", "UB6", "UB10", "UB12", "UB13", "UB16", "UB17", "UC1", "UC3", "UC5", "UC6", "UC7", "UC9", "UC11".[89]

c) In the Mediterranean Sea (Pola Cattaro) nine boats – "U21", "U33", "U34", "U35", "U39", "UB1", "UB15", "UC12", "UC14".[90]

d) In Constantinople five boats – "UB7", "UB8", "UB14", "UC13", "UC15".

e) In the Baltic Sea (Kurland Submarine Flotilla) five boats – "UA", "U9", "U10"," U17", "UC4"[91].

Sum 44 boats.

Looking at these numbers, doubts might arise as to whether the Baltic Sea and Constantinople deserved the allocation of so many boats compared to the total number, since the boats there apparently took away from the main purpose of the submarine war-forcing England to make peace by destroying their merchant tonnage and cutting their supply lines. To respond:

The battle for the Dardanelles was the focus at the time. If the Dardanelles front were defeated by the Entente we would have lost the war. It was therefore urgently necessary to send 5, sometimes 7 boats to Constantinople in order to take action against the enemy armed forces in front of the Dardanelles and against the Russian water routes between the Crimea and the Caucasus coast. The C-boats were used to transfer war material to the Dardanelles.

The Baltic Sea was only a secondary theater. Which is why the large boats "U23", "U25" and "U26" had been removed from there in the period described. The recall of "U26" to the Baltic which unfortunately resulted in the loss of the boat, was hard felt. There

[89] Footnote in original. As mentioned above "UB4" and "UC2" had been lost.

[90] Footnote in original. Boats "UB1" and "UB15" in July were handed over to the Austrian fleet July and June; "UB3" was lost.

[91] Footnote in original. Listed here because it was enroute. Started work at Kurland Submarine Flotilla on October 8, 1915.

can be no objection to the use of the other boats in the "Kurland" flotilla, which are listed in the above list under "Baltic Sea", because almost all of these boats were unusable in the North Sea war theater due to their technical condition.[92]

Boats complete or under construction:
a) Boats in service:
5 large boats: “U66”, “U67”, “U68”, “U69”, “U70” }[93]

---

[92] Footnote in original. Submarine activity in the Baltic Sea was too little related to general war activity and not extensive enough to commemorate it every month. However, a lot of effort and work has gone into the Kurland Submarine Flotilla (Libau base), and so the events are briefly summarized here.

"U4" under Oblt. Bräutigam (otherwise Submarine School) took part in a mining operation by the Commander of Baltic Sea Reconnaissance Forces from May 12 to 19, 1915. On May 14, "U4" luckily escaped the torpedo shot of an enemy submarine.

"UA" led by Oblt. Wilhelms made eleven and under Oblt. Scherb three cruises, most of which the boat provided reconnaissance for our naval forces operating off the Russian coast. Under Wilhelm it was active from August 1st to 5th in the Gulf of Riga, where it had various encounters with Russian destroyers. On July 4, a torpedo double shot by an enemy submarine was successfully avoided.

"UC4" under Oblt. Vesper laid a mine field at Hekla Lileina (west of Bengskär) on August 4; on August 14 at Utö. On August 12, it was also unsuccessfully attacked by an enemy submarine with two torpedo shots. On September 13, it laid a mine field at the north entrance to Moonsund.

"U26" had already sunk a steamer in the Gulf of Finland in April in addition to the minelayer (June 4). On August 11, 1915, it ended up taking a patrol position in front of the Gulf of Finland together with "U9".

"U9" (Spieß)) sank a steamer off Moon Island after several skirmishes with enemy submarines; from October 2nd to 10th, it guarded the entrance to the Gulf of Finland and Utö.

On August 7, "U10" (Stuhr) fired artillery on the Bengskär, had battles with enemy steamers and the Odensholm lighthouse. From September 22nd to 28th, it guarded the entrance to the Finnish Gulf to attack enemy submarines.

English submarines entered the Baltic Sea and created considerable trouble due to successful attacks on our Baltic cruisers and German commercial steamers.

[93] Footnote in original. These eight boats were, apart from the trials that had to do them, available to the Submarine School.

This included the boats "U1", "U2", "U3", "U4" and "U25". The latter boat was assigned there at the end of August, but was still under repair until April 1916 at Kiel shipyards. "U 16" started work at the Submarine School in mid-October 1915.

3 small boats: “UB9”, “UB11”, “UC8”. }90

b) Boats under construction:
In addition to the boats already mentioned building in volume 1, new construction was as follows:

In June 1915: The large boats "U81" to "U86" at the Germania Werft, Kiel, "U87" to "U91" from the Imperial Shipyard in Danzig.

In August 1915: The large boat "U92" at the Imperial Shipyard Danizig.

In September 1915: The large boats "U93" to "U98" at the Germania shipyard, the large boats "U99" to "U104" at the Weser shipyard in Bremen.

A total of 55 large boats were under construction. In addition to the 94 BII and CII submarines mentioned being built in volume 1, other small boats are not in this section.

## Final Remarks

In retrospect, when we saw the "submarine blockade in the waters around Britain and Ireland" collapse, we recognized as causes the premature onset and "politics" deprived the blockade of the full severity compared to what it could have done.

It should be emphasized here the submarines, their commanders and their crews have in no way and nowhere failed. The performances of the submarines were brilliant in this second phase of the war, despite all the bitterness created among commanders and crews by their political leadership’s attitude towards submarine warfare.

We have met the pioneers who showed our boats the way into the vastness of the Atlantic and into the most hidden hideaways of the Mediterranean. We saw individual boats rise to extraordinary, never before seen performances. The unreserved admiration of the whole world, including even our enemies, was given to the gray steel hulls with their crews, who also seemed to have steel hearts. The pride and self-confidence of submariners grew steadily. No one doubted victory must be ours with consistent political and operational leadership and with correct, energetic organization of all personnel and material, which had to be utilized more often in accordance with the extraordinary widening of the submarine war.

The loss of brave comrades and excellent commanders was lamented, but everywhere there was striving with willing fulfillment of duty and with full commitment of life and limb to do the same to those who had sealed with death their loyalty to the Emperor and the Empire and their love for the Fatherland.

# German Submarines and their Warfare
# 1914-1918

## Lieutenant-Commander Albert Gayer

### Volume III

### October 1915 to the end of April 1916
### The Winter of 1915/16 and the Period until the Order for Submarine Cruiser Warfare

# Part III - The period from October 1915 to the end of April 1916

# Forward

In order not to make the description too broad, an exact list of the individual vessels sunk has generally been dispensed with. Where losses are specified to the nearest ton, these are historically correct results obtained from the white papers of the countries in question, to which the sunken ships belonged.

Unfortunately, such white papers from France, the United States, Greece, Japan and Russia were not yet available, so some sinkings have not yet been historically confirmed.

I hereby express my most sincere thanks for the support granted to KKpt. Otto Groos from the Naval War Archives, who subjected both this and the second volume to a critical review and corrected various errors in the text, and KptLts. Wassner and Steinbrinck, who reviewed what was stated regarding Flanders, and finally General Staff Captain Geyer, who gave me various valuable tips,.

Berlin, February 1921 The Author

# PART III - The period from October 1915 until the end of April 1916

## General characteristics of submarine warfare in this period

The foreword to volume 1 stated the period now to be described did not express any clearly defined operational idea for the overall submarine war:

This should be discussed in more detail before the individual events are described.

Various goals were pursued in the individual main war zones of the North Sea, Flanders and the Mediterranean:

We saw that, according to the Kaiser's Decree of September 18, 1915, any kind of trade war on the west coast of Great Britain and in the Channel had to be suspended.

In the North Sea, the operations area of the "Commander of Submarines", the command was taken literally and the boats were used purely for military purposes. It was similar in Flanders, except mine boats could continue to work there, causing further damage to enemy trade and military transports across the Channel.

In order to be able to take effective action against the latter, the Flanders Submarine Flotilla was again permitted by an order dated November 21, 1915, to destroyed without warning enemy flagged ships, apparently heading to or fro French harbors between Dunkirk and Le Havre (including both places),.

Hospital ships, passenger steamers and neutral steamers were excluded.

The submarine trade war continued undisturbed in the Mediterranean. There it could lead to less incidents with neutrals, due to the almost completely lack of armament on steamers and the lack of trade routes security, the submarine war could be waged as legitimate cruiser warfare. The first incident was the Ancona Case in November 1915, but the subsequent exchange of notes between Austria-Hungary and the United States had no significant operational impact on the submarine trade war in the Mediterranean.[94]

When, at the end of April 1916, the naval war leadership ordered trade war to be conducted according to internationally recognized Prize Rules, the Commander of Mediterranean Submarines accepted it there without objection and found it quite feasible, while the Commanders of the North Sea and Flanders Submarine Flotillas raised the sharpest objections. This clearly shows the differences among operational capabilities for boats operating in each of the three main theaters.

But even in the period under review, the submarine trade war did not cease in the North Sea and Flanders. Rather, the situation underwent the following development: At the beginning of December, the Naval Staff feared submarine operations in the Mediterranean, which had been very successful and almost without loss, would soon be severely handicapped due to anti-submarine forces sent from England to the Mediterranean (patrol vessels, submarines, submarine traps, enemy submarines), if North Sea submarines were no longer active on the British coast; only the work of Flanders submarines held part of the patrols in the Channel.

It was therefore decided to send an experienced submarine to the English West Coast and "U24" — KptLt. Schneider — was chosen for this task. On December 16, 1915, the boat sailed from Helgoland

[94] The German submarine "U38" attacked the "Ancona" while flying the Austro-Hungarian flag. (Italy not being at war with Germany at that time.) Several U.S. citizens lost their lives in the sinking.

to go north around Britain to the West Coast. The boat received special orders not to sink passenger and neutral steamers unless they attacked "U24". We will return to this "winter shakedown cruise" to the English West Coast.

A further step towards the resumption of the submarine trade war was the entry of Bulgaria into the war as part of the Central Powers. The Chief of the Army General Staff, General v. Falkenhayn, considered the moment appropriate for recommencing submarine trade warfare, insofar as the danger of intervention by other neutral powers was considered less after the entry of Bulgaria. The Chief of the Naval Staff, v. Holtzendorff, strongly supported these efforts. Tirpitz campaigned for this by emphasizing the time had come for an intensified submarine war. In his opinion, boat quantity was sufficient and any further delay would bring only harm. Discussions between Army and Navy Command began on this topic on December 30, 1915 at the Ministry of War.

Two memoranda from the Admiralty Staff followed on January 7 and February 12, respectively. In these pamphlets, the expectation was expressed if the submarine was ruthlessly used English resistance would be broken in at most half a year.

Risk of American involvement was taken into account, but it was emphasized the war would have to end, if possible, by autumn 1916, otherwise the hope for a peace settlement beneficial to Germany would wane. The second memorandum also went to a large number of economic experts who unanimously expressed their approval regarding the immediate resumption of unrestricted submarine warfare as the last opportunity for Germany.[95] As early as February 24, 1916, the order was given to recommence the trade war provisionally according to the Prize Rules and with the proviso that vessels identified as troop transports and armed merchant ships could be sunk without warning. It did not matter whether the

---

95 "Die Englische Wirtschaft und der U-Bootskrieg" consisting of 38pp. of text and a 66pp. statistical supplement.

steamer was known to be armed according to official declarations, etc., or if the submarine identified armament being mounted.

Passenger steamers should never be sunk, even if they were armed. In March, it was then ordered that enemy merchant ships within the war zone could be destroyed. However, this provision was not to remain long in effect:

On April 24, due to the sinking of the passenger liner "Sussex" by the Flanders submarine "UB29"– the commander had identified the ship as a troop transporter seeing there were English troops in uniform on deck[96] (March 24, 2016) – the order was issued for submarines to only wage trade warfare according to Prize Rules until further notice.

We shall address later the consequences of this command, which ushered in a new era.

A decisive meeting occurred on March 6, 1916, in Wilhelmshaven; the Emperor rejected the wishes of General von Falkenhayn to expand unrestricted submarine warfare to include attacking without warning all vessels within the blockade area including neutrals and sparing only hospital ships and the like. Von Tirpitz, who was a proponent of intensifying submarine warfare, was already removed from affairs and not invited to the meeting, resigning on March 17th. History has proven him correct. In March it was high time to declare unlimited submarine warfare. Even if America declared war at this time, there is no doubt England, with all its anti-submarine development, was far behind the level attained in February 1, 1917, when unrestricted submarine warfare was declared. Armed steamers were only available in small numbers. A great deal of time and effort was required to create the 20,000 guns we calculated where needed to arm the larger enemy steamers!

[96] The troops on deck were actually civilians as Sussex was still being used for civilian traffic; exemplifying the difficulty of accurate identification via a periscope.

Jellicoe's efforts, initiated in summer 1915, to create dense mine fields in the German Bight, failed because a useful type of mine was not yet available; it was not developed until 1916, after which 100,000 of these new mines were ordered. However, as recognized by Jellicoe, they would not produce results until late 1917; by then we would most likely have won an intensified submarine war begun in March 1916. Hundreds of contracted patrol and escort vehicles were still in the first stages of construction, and in 1917 they were ready.

Another question is whether there were enough boats to start unrestricted submarine warfare in March 1916. Let us consider the number of boats in the main war theaters in March 1916:

## North Sea.

After the overhaul that had taken place in winter, the following boats were now fully ready for operations:

"U19", "U20", "U22", "U24", "U28", "U32", "U43", "U44", "U45", "U66", "U67", "U68", "U69 "," U70 "," U73 "," U74 ", a total of 16 boats.

Boats "U46" to "U50", "U71", "U72" and "U75" to "U80", "U51" to "U65"[97] were expected to come into operation by late summer, so 16 large boats ready and 28 large boats almost completed; in addition to these boats, the small boats "UB 21", UB 22 "and" UB 27 "joined the North Sea submarines in late April.

## Flanders.

"UB2", "UB6", "UB10", "UB12", "UB13", "UB16", "UB17", "UB18", "UB19", "UB26", "UB29", "UC1", "UC3", "UC4", "UC5", "UC6", "UC7", "UC10", a total of 18 boats.

Reinforcements consisting of the entire BII and CII series boats up to "UB48" and "UC79"were almost ready in 1916 and during the

[97] Footnote in original. The rapid completion of boats "U63" to "U65" contracted with Germania shipyard in early 1915 is particularly remarkable and is due to the use of motors ordered prewar for Russian submarines.

unrestricted submarine war would be sent to Flanders, as they were not used in the Mediterranean.

The reason for using high quality boats such as six BII and a CII boat in the Baltic Sea in 1916 instead of Flanders was due to unrestricted submarine warfare not coming to fruition that year.

**Mediterranean Sea.**

"U21", "U33", "U34", "U35", "U38", "U39", "UC14", 7 boats.

Assigned to the Mediterranean immediately after completion: “UC 20”, “UC 22”, “UC23”, “UC24”, “UC34”, “UC35”, “UC37”, “UC38”, “UC52”, “UC53”, “UC54”, “UC67”, “UC73”, "UC74”, “UB42”, “UB43”, “UB44”, “UB45”, “UB46”, “UB47”, i.e. 20 boats.

**Constantinople.**

"UB7", "UB8", "UB14", "UC15", "U33", i.e. 5 boats.

**Baltic Sea**

"UA", "U9", "U10", "UB 2", "UB 5", "UB 20", i.e. 6 boats.

This list demonstrates there was a completely sufficient number of boats and a strong reserve for waging aggressive unrestricted submarine war. The number of available boats was certainly higher on February 1, 1917, but the effectiveness of individual submarines on the enemy was at least twice as high in 1916 as in the 1917 average. Proof of this position will be given later. It is based on the already mentioned much less developed 1916 enemy anti-submarine measures. A comparison of the above data with the number of boats available in the initial 1915 submarine blockade can be seen as follows:

In February 1915
At the front. 26 large and 1 small boat
Under construction. 42 large and 127 small boats.

In March 1916
At the front. 25 large and 43 small boats.
Under construction (almost finished). 52 large and 89 small boats.

When making this comparison, it must be borne in mind that among the 26 large boats with which we entered the submarine blockade in February 1915, 7 were petrol boats and therefore inefficient. The 25 large boats available in March 16 belonged to the newest type (except for three), had been battle-tested and combat experienced crews. With a few exceptions, the boats listed "under construction" in 1915 were still very far from completion. The 94 BII and CII boats only started to arrive at the front in February 1916. The boats that were referred to as "under construction, almost finished" in March 1916 arrived in April with unprecedented speed. During the war, the monthly reinforcement of new boats in service never got as low as it did for the period from April 1, 1916 to January 1, 1917.

In 1916 came into operation on a monthly basis:*

| | |
|---|---|
| March | 9 boats |
| April | 8 boats |
| May | 9 boats |
| June | 8 boats |
| July | 9 boats |
| August | 10 boats |
| September | 10 boats |
| October | 11 boats |
| November | 11 boats |
| December | 15 boats |

*A more favorable monthly increase would not materialize until November 1918 after the efforts of the Submarine Office. From November 1918 to November 1919, the average monthly growth would have been about 27 boats, including "large" and "extra large" models.

While there were differences, as highlighted above, regarding operations in individual theaters of war were, the operative principles for boat at the front outlined in volume 2, pp, 37 to 41, also proved their worth in this third phase of the war, they increasingly became standard procedure for all submarines. It was very satisfying for the submarine flotilla commanders, who had laid the foundations with the assistance of most experienced submarine commanders. The English press noted when the trade war reopened in March 1916: German submarines have learned a great deal and are unlikely to be fooled by clumsy ruses as they were in 1915. Especially recognized as an important operating principle by the boats - The English in their to some extent sporty view of naval warfare recognized the qualities of prowess in their enemy - was reported in the enemy press that submarines now either submerged immediately after firing warning shots at a steamer to force the crew to abandon ship or approached the steamer surfaced and with continuous artillery fire to force any hidden gun crew onboard ship to reveal themselves. If the latter procedure was sometimes maliciously referred to as "firing on the lifeboats", it was a lie, the English trade press is rarely of the same opinion and almost always recognized the military purpose of this procedure.

Both the preliminary training cruises of new submarine commanders with experienced commanders as well as the the aforementioned system of war pilots continued to prove themselves very well.[98]

Boat losses remained very low in this 3rd phase of the war. They averaged only one boat a month. The procedure for large North Sea submarines to travel northward to their operating areas on the west coast Great Britain was retained, which had a favorable effect on the loss rate. The senior command approved the careful entry of boats before commencing operations; new boats were gradually introduced into warlike conditions by simple trips. Only "U68" (KptLt. Ludwig Güntzel) was lost in the North Sea.

---

[98] Footnote in original. See 2nd volume, page 38.

From the summer of 1915 onwards, the activity of boats in Flanders was hampered by, in addition to the formerly inadequate and gradually improved mine barrage in the Dover Straits (see volume 2, page 42), the enemy had a large net boom across the Channel (at the height of South Goodwin Sands beginning and extending from about 110°). This boom hung on large floats, buoys, and rafts; one had to pass them surfaced at night; because submerged, the boats could get caught in the net very easily. However, since the enemy had to take precautions for their own ships' passage, he was required to lay out various light buoys, which were not only distributed around the net itself, but in some cases also at a greater distance from it and served as warning and navigation buoys. These buoys soon became trusted friends of the Flanders flotilla. They received special numbers and were an excellent navigational guide for passing Dover Strait. Even unlit large buoys, which the enemy had to lay out to secure his own navigation, were very advantageous for us; a large lattice barrel at 51°20.6' N and 20° 7.2' East was for a long time the steering point for all submarines wanting to pass the Dover-Calais narrows from the North, it was labeled 2501. Close to it, lying on the bottom, the boat usually waited for a favorable time of day and current to pass the Strait. The cruises of the small Flanders boats in this period had already become "dashing hussar raids" deserving of great admiration. Flanders theater became, in all fairness, a high example of the submariner's art. It was astonishing to see what kind of danger the small, maneuverable submerged boats almost always escaped under resolute command. Flanders commanders produced these achievements through extensive self-training and instruction, In this period, only "UC9", "UC5" and "UB26" were lost in Flanders, "UC8" was interned in Holland after going aground at Terschelling. At the beginning of 1916 enemy anti-submarine measures in Flanders became stronger and more consistent. This was especially true regarding the C-boat minelaying areas and later in April 1916 (on Easter Sunday) laying out a large network of mines along the entire Flanders coast at a distance of about 15 nm; also to the approach routes of all Flanders boats, in April 1916 (on Easter Sunday) laying out a large network of nets and mines along the entire Flanders coast at a distance covering about

15 nm Aircraft and patrol boats protecting this net tried to force submarines to submerge and chase them into the net. Enemy aviators in Flanders were an increasing nuisance to our boats.

Work was relatively easy in the Mediterranean at the end of 1915. Arming of steamers was scanty as well as trade route protection being insufficient. The Entente simply did not have enough small vessels to implement a similarly strong security protection in the Mediterranean Sea as around England. On the other hand, working in front of the Dardanelles and Salonika was difficult; there the enemy had set up the same patrol protection as at home. At the end of 1915 there was also some fishing trawlers and destroyers guarding Otranto Strait. However since it was not British directed it was not particularly effective for the time being. Patrolling steamship routes, escorting valuable vessels and arming steamers operating in the Mediterranean increased slowly and steadily during the third phase of the war. At the beginning of 1916, it sometimes proved to be a hindrance (see “U35” March cruise, page 146). What particularly facilitated submarine activity in the Mediterranean compared to the northern operating areas was the significantly better visibility and easing of swells soon after winds had reduced, in short, the weather was much favorable on average. Every submarine commander going to the Mediterranean was much envied by comrades who stayed at home because of the extraordinary operational possibilities in the Mediterranean. The performance of the Mediterranean boats in this period should not be diminished in any way by these explanations. On the contrary, the work of all submarine commanders there, almost without exception, deserves the greatest admiration.

During the 3rd war phase only "UC12" (commander Oblt. Fröhner) was lost in the Mediterranean.

Further boat technical improvement, their weapons and facilities important for warfare continued to grow, which we will see in the description.

In the third phase of the war occured the second change of the North Sea Fleet Commander; Admiral v. Pohl fell seriously ill in early January 1916. Admiral Scheer was asked to replace him temporarily, and on January 15, 1916 Scheer was confirmed in office. At the end of April 1916, as an opponent of any half measures and, in his opinion the current aimless use of submarines, Scheer rejected the submarine trade war according to Prize Rules and delivered his definitive verdict to the Naval Staff: Either unrestricted submarine warfare or exclusive use of the boats against warships. Since the former was rejected by the Naval Staff on the basis of the Kaiser's decision, the fourth phase of the war from May to September 1916[99] actually brought about a complete cessation of the trade war in the North Sea. At the end of this volume there will be a detailed discussion of this momentous decision.

## September and October 1915

### North Sea

If from September 20, 1915 no more boats were assigned to trade warfaer, this did not mean that there was now a rest at the North Sea bases; on the contrary, the onset of the purely military activity of the boats brought a lot of work.

On September 24, "U28" and "U43" were positioned in front of the Firth of Forth and Pentland Firth, respectively, to stalk enemy warships. Both boats were out of luck. On the way north before reaching the Norwegian coast, "U28" became very nose heavy and unable to maintain trim such that the integrity of the battery could no longer be guaranteed. It had to return to Wilhelmshaven. "U43" suffered a fracture of the forward diving planes and had to return to the Ems. These accidents confirm the statement made on volume 2

[99] Footnote in original. * The order for 3 boats to be sent to the northern Arctic Ocean and conduct trade warfare according to Prize Orders was issued on September 15, 1916; As a single great stroke, Fleet and MarineKorps Flanders were ordered on October 6 to simultaneously commence trade warfare under Prize Orders everywhere on October 15, 1916; at the same time "U32", "U63", "U64", "U65" was ordered to the Mediterranean.

page 85 that by no means were all large North Sea boats suitable for use on the British west coast in winter weather conditions.

Due to the intended use of the boats against warships, their training in long range torpedo shooting was promoted as often as possible. Appropriate exercises were held at Borkum-Reede.

Mid-October weather conditions favored possible enemy operations. Patrols were strengthened at the mouth of the Ems and Borkum-Reede by one or more submarines. "U44" and "U19" were on outpost duty at Horns Reef at that time.

On October 20, “U38” (KptLt. Valentiner) left for the Mediterranean. Valentiner would be another commander sent to this operations area who would have extraordinary success there. Immediately after entering the Mediterranean Sea, he sank 2 troop transport steamers, 11 steamers and 1 sailing ship on the way between Gibraltar and Malta (3 to 9 November). The signal station at Cap de Fer lighthouse in Algiers tried to warn approaching steamers of the submarine. It was taken under heavy artillery. This cruise provides a fresh interlude compared to the tedious, futile activity against warships, which at that time North Sea boats were engaged in.

On October 20, "U19" and "U32" departed simultaneously from the Ems. "U19" was supposed to stalk enemy warships in front of Scapa Flow, "U32" in front of Peterhead. However, both boats went in the evening to Lister Deep due to bad weather, where they were then held by the fleet for a planned operation. The next day, "U44" also ran into Lister Deep because of an east storm.

On October 22, 1915, "U67", the first of the new Germania boats (ex-Austrian) arrived at the Ems, commander KptLt. v. Rosenberg-Gruszezynski, whom we met earlier as commanders of "U30".

On October 23, "U32" and "U44" were ordered to undertake an operation with the fleet on October 24 and occupy certain positions north of Horn's Reef. Because of the ensuing bad weather and a number of other adverse circumstances the fleet soon turned back and the boats did not even come out.

"U19", which departed on October 23rd enroute to Scapa Flow, suffered a malfunction in an electric motor on October 24th and therefore returned to Wilhelmshaven.

"U32's" trip to Peterhead did not amount to anything, to which the boat departed to on the 24th. Due to a hurricane-like storm the boat came back in on October 26th. "U44" was supposed to take a position north of Horns Reef and returned for the same reason.

October in the North Sea ended with the failure of many intentions due to a confluence of unfortunate circumstances.

### Mediterranean Sea

October brought a number of great successes, which totaled 75,000 tons of enemy shipping capacity, while in the North Sea and Flanders there were no torpedo sinkings at all.

At the end of September, we saw KptLt. Forstmann successfully start his first Mediterranean cruise, in the area between Crete and the African coast to Cyprus. In October, he sunk 4 more cargo steamers in addition to the 3 already sunk at the end of September. Working on the steamer route from Port-Said to the West proved to be extremely productive. On October 12, he carried out an attack on an outpost line off the Montenegrin coast consisting of 5 steamers; one of the steamers was sunk so that the total was 22,287 tons.

"U33" under KptLt. Gansser was, from September 28 to October 9, 1915, is working the Cerigotto Strait- Crete passage. On October 1 and 2, he sank the French auxiliary ships "Provincia" and "Ste. Marguerite"; also during the cruise sank 2 troop transport steamers, 7 steamers and a sailing ship, together about 50,000 tons. Gansser, a native son of Württemberg, went into Mediterranean operations with tremendous zest like Kophamel, Forstmann and Valentiner. If his overall performance was less than, for example, Forstmann and

Valentiner who flourished there, the reason for this is "U33" based from Constantinople, operated in the Black Sea throughout 1916, and the possibility of great success there was much less than in other areas. It was of course impractical to keep Gansser with his powerful boat away from the main theater of operations and instead in the Black Sea- a secondary theater. BI and BII boats would have been sufficient for the activity there. Throughout 1916, Gansser was only able to sink 4 steamers, 4 sailing vessels and 2 fishing cutters in the Black Sea apart from successfully bombarding some enemy coastal sites; in the Mediterranean he could have achieved at least ten times the result in the same period.

"U35" under KKpt. Kophamel was underway on a long cruise from October 12 to November 12, 1915. Initially he went to the Aegean Sea to disrupt troop transports traffic on the routes to Saloniki and destroying enemy supply ships. From 16 to 27 October, Kophamel operated in the Gulf of Thessaloniki. A smaller steamer and the English transport "Marquette" displacing 7,057 Br. Reg. Tons were sunk during this period; the latter contained 1,000 white troops, of whom 29 were rescued[100], as well as a large ammunition load on board. After this sinking, transport traffic from Salonika took place almost exclusively at night, and Kophamel now went to the "Hersing booth" in the Gulf of Saros to received messages and orders. In addition, there was a rendezous date to embark a Turkish mission to transport to the Gulf of Kos (meeting point Orak). Kophamel's further adventurous and successful cruises will be described next month.

"UC14" made a cruise from Constantinople to Varna in early October.

When the Bulgarian port of Varna was attacked by Russian armed forces and aircraft on October 27, "UB7" under Oblt. Werner went on the attack. A hit was achieved against an enemy warship. "UB8"

[100] Actually 167 perished of 741 onboard in the sinking.29 of the crew perished which may account for the juxtaposition of information.

also participated defensively against this attack, without being able to achieve any success.

Successful as always, "UB14" under Oblt. von Heimburg was cruising in the western part of the Black Sea from October 3rd to 19th. A Russian steamer was sunk on October 7, and another on October 8, which had previously unsuccessfully attempted to ram "UB14".

## Flanders

In Flanders, as in the North Sea, the order to end submarine warfare initiated a period of stagnation, especially for the UB boats (torpedo submarines). The mine boats operated with all the greater zeal.

"UC1" (Oblt. Werner), "UC13" (Oblt. Wassner), "UC5" (Oblt. Pustkuchen), "UC6" (Oblt. Pustkuchen, Graf v. Schmettow), "UC7" (Oblt. Wäger), "UC9" (Oblt. Schürmann) were responsible for activity in October. Unless the boats were prevented by an occasional long yard overhaul, they tirelessly carried one mine load after the other to the enemy coast. Pustkuchen and Graf Schmettow were particularly busy; the latter e.g. 5 trips this month; the former reached Portsmouth and disrupted both access routes there.[101]

These trips usually led close to enemy controlled waters and were not without peril. "UC6" was rammed by an enemy torpedo boat in early October, but got off with slight damage. In mid-October the boat got into nets northwest of Calais and had to cancel its cruise due to an unclear screw.

"UC7" got into a fishnet in front of the Humber on October 20th, from which it was able to free itself.

[101] Footnote in original. The success of all these individual cruises cannot yet be precisely determined, since enemy material has not yet been collected. The researcher will find this a worthwhile undertaking.

"UC9" under Oblt. Schürmann, who had just arrived in Flanders, set sail on October 20, 1915 to south of the Longsand Lightship and did not return; nothing is known about the cause of the loss.

The existing white papers cite the successes in mid-October, which can certainly be attributed to the activity of the Flanders mine boats: 6 merchant steamers destroyed for a total of 12,546 tons. Some of the losses occurred close to the enemy coast, e.g. in front of the Folkestone Pier and under South Foreland. In addition, 3 English steamers of 6,564 tons ran onto mines, but could be saved.

3 Norwegians and 1 Dutch steamer were lost; an English destroyer ("Velox"), an admiralty yacht and 6 minesweepers also sank.

As mentioned, these successes are not extracted from the submarine success reports, but are taken from the white papers published since the end of the war. They do not yet give a complete picture, as only English, Scandinavian and Dutch sources were available for compilation; however, they are sufficient to show what disturbances the 6 primitive boats caused in their sphere of activity. A result of around 20,000 tons this month could be attributed to them.

The small B-boats were no less tirelessly active in Flanders, even if, despite the restrictions imposed on them, they were unable to achieve sinking successes.

"UB17" under Oblt. Wenninger was on observation post at Middelkerke-Bank from October 2nd to 4th. During the night of October 2nd to 3rd, the boat sighted 8 shallow draft enemy vessels approaching the coast to fire at Zeebrugge. Wenninger reported an alarm in time so the entire coast could be alerted and the approaching enemy repulsed; he approached in the early morning and shot his two torpedoes, which had passed beneath them due to the vessels' shallow draft.

The increased outpost service by B-boats continued throughout the month, as it was feared there would be hostile landing attempts in connection with the offensive underway at Loos.

"UB5" under Oblt. Smiths was dispatched to the Baltic Sea on October 6 upon reception of the Naval Staff's order to provide the Baltic Sea Commander with an available submarine.

"UC11" under Oblt. Walter Schmidt went to Kiel on October 17th to be exchanged for "UC8" belonging to the submarine school.

## November 1915

## North Sea

The weather conditions in the North Sea made this month's tasks extremely difficult. Most of the boats in the North Sea at that time were very sensitive to wave impacts on the hull on and especially the hydroplanes. It was therefore not allowed to expose them to heavy weather right away.

A genuine Meteorological Service was set up at this time at the instigation of the Submarine Command. So far, there had been hardly any manner to obtain a weather overview of the North Sea, let alone obtain weather reports for the submarine operations areas. Unfortunately, obtaining barometer readings from the enemy coast (for example through agent reports) proved impossible. The Submarine Command now pressed that weather reports at least from Denmark and Norway were to be transmitted daily by the Hamburg Naval Observatory. If one knew the barometer and wind from the area of Bergen (Norway)[102], Skagen, List, Helgoland and Borkum, one could at least have an idea of North Sea weather conditions and recommend the best boat routes, i.e. Depending on the pressure distribution, either a route off the Norwegian or the

[102] Footnote in original. Attempts to obtain weather reports from Iceland were unsuccessful.

English coast.[103] One continued to get an idea for the choice of the right departure time. It was recognized from the weather reports for example, if only in the German Bight there was bad weather, but there was higher pressure off Norway, it was then possible to let the boats push through the bad weather in the German Bight; on the other hand, the weather reports occasionally gave reason to hold boats back in the German Bight despite favorable weather there because there was heavy weather in the northern North Sea. The latter case is particularly common in southern and southwesterly winds. These winds are usually less strong in the southern North Sea in winter, while their force increases to the north of the Dogger Bank.

The British were naturally vastly superior to us in the possibility of correct weather assessment in the North Sea. The telegraph stations of all countries bordering the North Sea (except Germany) were available to him at any time to get weather reports. This explains the extraordinary precision with which the English armed forces »including enemy submarines« - were able to stand towards the German Bight with good weather. They sailed just after receiving news of expectant good weather. (Sailing only after good weather has occurred is usually too late in the North Sea if you intend to perform an operation in winter).

As a result of suspected mines in the sea area west of the Ems off the Dutch coast, submarines had become accustomed to traveling north via Helgoland, west of the Amrum Bank, Horns Reef, so in the second half of November a cleared mine passage was started at the Ems Hubert Gat using the following route: W½ by N 24 nm miles from the Hubert-Gat control buoy, then NW5/8 by W35 nm; this route was cleared to a width of 2 nm. This route is described so precisely because it would be used for almost a year for ingress and egress by Ems submarines without any interference from the enemy. Ameland Lighthouse which was the starting point of the

[103] Footnote in original. The weather service was supplemented by weather reports submarines had to submit on their way to and from the west coast, even in the northern North Sea.

NW course, offered a very good navigation point for incoming boats. It was not until December 1916 that the enemy disturbed this rout with mines. (See Fig. 14[104])

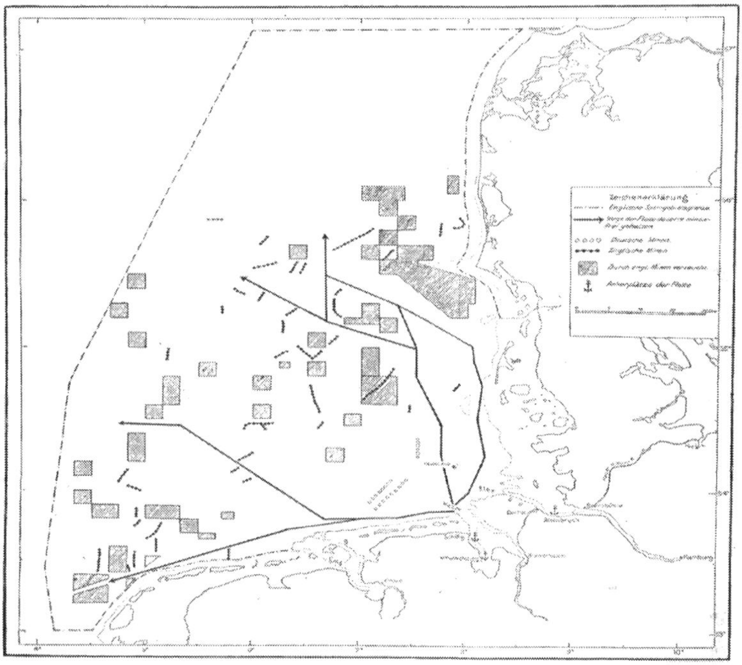

Fig 14. Solid lines are German submarine routes, boxes are German minefields, dotted lines are British minefields.

In November was tested the underwater signaling system which had been installed in the III Submarine Flotilla Leader Boat,"T137". Valuable experience was gained regarding submerged submarine signal traffic with surface vessels. This signaling method soon became indispensable for the purposes of submerged boat exercises.

At the end of the month, the lifting ship "Oberelbe" from the "Nordic Salvage Association" was permanently posted to the Ems, together with the tugs "Möwe" and "Seeadler". With the number of submarines increasing rapidly it was considered important to have

---

[104] Reinhard Scheer, "Deutschlands Hochseeflotte im Weltkrieg. Persönliche Erinnerungen." Scherl, Berlin 1920.

a powerful lifting vessel on hand that could intervene immediately in the event of an accident. Fortunately, "Oberelbe" was not needed for such duties at the Ems, while the two tugs soon became indispensable for transport trips, towing targets and submarines, and as an icebreaker in winter. The brave captains Giese and Käselau of these vessels became well-known supporting submarine operations n the Ems.

Among the most prominent operations were:

“U19”, from November 15th to 27th, traveled to the east side of the Orkneys to investigate the eastern entrances to Scapa Flow. Breaking through the outlying chain of patrol steamers, it went close inshore between Pentland Skerries and Copinsay Island.

"U20", from November 16 to 23, surveyed for a mining operation by the auxiliary cruiser " Möwe ", the north coast of Scotland to west of the Orkneys, to the north entrance of the Minch and in particular the area near Cape Wrath, regarding lighted beacons, patrols, the presence of enemy forces as well as ordinary navigation in this area. War Pilot Badewitz had the opportunity to familiarize himself with the circumstances. After his return he transferred to the auxiliary cruiser "Möwe" (commander KKpt. Graf zu Dohna) and did valuable work in the operation.

“U28”, Freiherr v. Forstner, and "U67", v. Rosenberg-Gruszczvnski, were in the Skagerrak region from November 16 to 24 to attack enemy warships. There was no encounter with the enemy.

"U43" (Jürst) observed English warship traffic south of Peterhead from November 17th to 26th. Nothing was seen by the enemy.

In addition to individual cruises, the boats were used in outpost positions west of the Ems and north of Horns Reef.

As of November 9, the I Submarine Half-Flotilla’s Commander’s, Staff and Leader boats were used until further notice by the

Commander of Baltic Sea Strike Forces, the boats were otherwise distributed.

The Flotilla Commander, KptLt. Pasquay, soon began operations there.

**Flanders**

The operational picture was generally the same as in October: strong activity of the mining boats, B-boats for outpost duty, which was loosened after a decision on November 5 because an intended landing of the English on the Flanders coast became less likely. This freed up some B-boats for long-distance cruises. The war conduct order mentioned in the introduction that shipping between the French ports of Dunkirk and Le Havre could be torpedoed again without warning opened up a new field of activity for the B-boats.

"UC1" (von Werner) planted 3 times, "UC3" (Wassner) 2 times, "UC5" (Pustkuchen 3 times, "UC6" (Graf v. Schmettow) 1 time, "UC7" (Wäger) 3 times their entire boat mine loads.

With the exception of two trips, all carried mines to the English coast (Thames estuary): Pustkuchen went through the Calais Narrows again at the end of the month and laid mines at Bassure de Bas; Count Schmettow on November 7th planted a field in front of Le Havre, on November 8th, one in front of Boulogue.

"UC8", which was released from Kiel to Flanders instead of "UC11" unfortunately went aground off Terschelling after leaving Flanders and was interned in Holland. The Dutch Navy later acquired it through purchase.

According to the “White Book”, the following results were achieved by Flanders mines in November: Warships sunk were 1 English minelayer and 2 mine sweepers, as well as 9 English merchant steamers totaling 9676 tons and 3 Norwegian steamers totaling 4,559 tons.

Of the remaining boats, the B-boats, this month Oblt. Weuniuger achieved success with "UB17". On November 9th, he sank a French torpedo boat near Bergues-Bank.[105]

## Mediterranean Sea

"U33" (Gansser) was from November 15th until December 6th in the Mediterranean to the west of Corsica. His total booty was 14 cargo steamers and 2 troop transport steamers; a huge haul of over 50,000 tons. It was a great loss to submarine warfare that this cruise concluded Gansser and his boat's operations in theMediterranean.[106] The great November success he achieved despite considerable disruption from heavy weather.

"U34" (Rücker) worked from November 3rd to 22nd between Crete and the African coast. A transport steamer was sunk, 4 cargo steamers (altogether 20,000 tons) sunk. Rücker first noticed a fishing trawler patrol in Otranto Street, dragging submarine nets. Since these nets did not yet carry explosives, they were harmless; patrol ships also appeared in Cerigo Street (passage between southern Greece and Crete).

We have mentioned the October cruise of "U35" (Kophamel) from Saloniki to Hersing's Stand. After everything had been arranged there, he went to the Gulf of Kos (meeting point Orak). The boat was here from October 31st to November 1st. The Turkish mission consisted of 10 officers and $13^{1/2}$ of cargo tons were taken on. On the evening of November 1, two sailing vessels loaded with ammunition and Turkish troops were taken in tow at Budrum Roads and the trek to Africa began. The English steamer "Woolwich" (2,963 tons) was sunk during the crossing. Arriving on November 4 at the port of Bardia (North Africa) where the cargo was unloaded quickly. On November 5, the English auxiliary cruiser "Tara" (1,862 tons) was sunk off Sollum. 14 rescued officers and 79 men were handed over

105 Wenniger reported the vessel as a "a great French torpedo boat" in the poor visibility. French records indicate it is an auxiliary patrol boat.

106 "U33" and Gansser were transferred to the Black Sea.

to General Nuri Pascha, who commanded in Africa. In the afternoon of the same day, two English gunboats in the port of Sollum were disabled by gunfire. Eight further large cargo steamers were sunk in the sea area south of Crete from November 6th to November 10th brought the result of this operation to almost 50,000 tons. It was an excellent performance by the brave crew covering almost 3400 nm in this most exhausting long-distance cruise, of which 410 nm were submerged. (Kophamel was one of the first sea officers to join the then "new" submarine service in 1909. When he put "U35" into commission on November 3, 1914, he was older than most submarine flotilla chiefs; nevertheless, he placed his extremely tough, no-nonsense, personality into the exhausting submarine service, where he showed great leadership.)

The 14 vessels sunk by Max Valentiner between November 3rd and 9th when entering the Mediterranean from Gibraltar to Malta are already mentioned in the October section. Their total tonnage was also around 50,000 tons.

“UC14”, Oblt. Bauer, on November 26th, laid a mine field in front of Valona, on which the Italian auxiliary cruiser "Citta di Messina" hit and sank on December 5th.[107]

"UB7", Oblt. Werner, sailed from Constantinople in the period from 19th to 24th November towards the Danube estuaries

"UB14", Oblt. v. Heimburg, who did not let any month pass without success, sank the English submarine "E20" on November 5 by torpedo fire in the Sea of Marmara.

"UC13", Oblt. Kirchner, on November 22nd in the northeastern Black Sea, forced a Russian steamship laden with oil to run aground by gunfire. He also sank 4 sailing ships. On the way back the little boat was caught a fierce storm. Being unable to make headway on the surface due to the weakness of the boat’s engines, an attempt was made to wait out the storm submerged, which was initially successful. After a lengthy period the boat was eventually forced to

107 The vessel survived this mining.

surface where it washed ashore in heavy seas near Kirpen and was lost; the whole crew was saved.

The Mediterranean alone had a sinking result of 176,393 tons in November, more than was previously achieved monthly in all war theaters. Together with the mine results of the Flanders boats almost 200,000 tons was the total for November, at a time when 9 large, powerful submarines were practically unused in the North Sea and the B-boats in Flanders could only be used to a very limited extent. Due to these results there was no shortage of voices at the front in favor of sending all suitable North Sea submarines to the Mediterranean.

For this purpose "U20", "U24", "U28", "U32", "U67", and after some improvements "U44" would have been considered.

However, due to the poor performance of the Mediterranean bases, the boats would have been withdrawn before long repairs were necessary. But they could have done two or more cruises. The commander of "U20", KptLt. Schwieger, strongly supported this idea. History must prove him correct that the boats mentioned could have achieved extraordinary performances in some medium-ranged operations. In the North Sea they lay idle until March 1916. "U19", "U22", "U43" and the boats that came to the front soon after would have been a deterrent to the enemy, especially since the Flanders B-boats could have gone to operations in the German Bight if necessary and reinforcements from the Baltic Sea could easily be fetched.

## December 1915

### North Sea

Although this month the North Sea submarines were not given such an opportunity like the boats in the Mediterranean, at least the year ended with the aforementioned cruise by KptLt. Schneider, "U24", to conduct trade warfare on the English west coast. Despite the most severe weather[108], which Schneider got into in various ways, he managed to sink 3 English and 1 Belgian steamer totaling 15,000 tons on the Irish south coast and off the Bristol Canal. The steamer "Huronian" of 8,766 tons, which Schneider hit with a torpedo, was still able to reach the protection of Bantry Bay.

"U20" was on outpost duty in the Skagerrak from 17th to 24th, "U44" from 14th to 25th in front of the Tyne mouth and the Pentland Firth. In the latter position it sighted 4 battleships on December 18 and on December 20 it shot with two torpedoes at an English small cruiser with four funnels; the boat could not get close enough due to the high speed and zigzagging courses and the torpedoes missed their target. The cruiser went into Scapa Flow.

New type of mines were laid in front of the Tyne estuary and "U44" delivered them out of the torpedo tubes. Training in laying minefields with this new weapon was done in the first half of December at the Ems.

Its commander, KptLt. Wagenführ, one of our oldest submarine officers, had thus started offensive operations in this theater of war. At the beginning of the war he was still an officer on board the gunboat "Lynx" in East Asia. Later watch officer on the auxiliary cruiser "Prinz Eitel Friedrich". With great difficulty he made his way to Germany in a Norwegian steamer as a stoker to serve in the submarine service.

"U67" was before Peterhead from December 18-27, the enemy was not sighted.

In addition to training "U43" and "U44" in launching mines via the torpedo tubes, a new ejection device for torpedoes was tried out in Borkum Reede in December, which was to avoid the strong water surge when firing. The treacherous bubble was significantly reduced

[108] Footnote in the original. The submarine's helmsman Heinrich, who had excellently supported Schneider on all his cruises, suffered serious injuries on December 17 due to heavy swamping breakers.

by the new equipment. Technology was directed towards completely eradicating any surface signs of a torpedo launch.

Through the efforts of the KKpt. Bindseil of the High Seas Fleet staff, the directional radio stations in List and Borkum were now so perfected that the boats could find their bearings to determine their ship's location when heading for the German Bight. This feature instilled much confidence in commanders regarding navigation safety - especially with the intensified enemy attempts later on to mine the bay.

At the beginning of December the new boat "U68" arrived at the Ems under the command of KptLt. Ludwig Güntzel. The boat went to sea on December 27 to provide reconnaissance services for the departing auxiliary cruiser "Möwe" (Commander KKpt. Gras zu Dohna-Schlodien). So at the end of the old year this boat helped to start an enterprise that was to be accompanied by great success and which anew increased respect for Germany's achievements at sea.

**Flanders**

Boat activity here was almost the same in December as in previous months. "UC1" - v. Werner – laid a minefield at the South Goodwin Lightship and three in front of the Thames estuary.

"UC3" - Wassner – laid a mine field in front of Boulogne, one in Black Deep, one east of Shipwash.

“UC5"- Pustkuchen planted a field in the channel leading to Harwich.

"UC6" - Count v. Schmettow - a barrier in front of the Humber, one in front of Calais, one at Dover; Working in front of the Humber meant a very special feat for the small boat at this time of year.

“UC7”- Oblt. Haag - planted 3 fields in front of the Thames Estuary, 1 in front of Stanford and Hewett Channel.[109]

The result in merchant ships was:
5 English steamers of around 9,500 tons sunk,
An English steamer of 4,844 tons ran onto a mine, but was salved,
4 Norwegian steamers of 4,176 tons sunk,
2 Swedish steamers dropped from 2,936 tons,
1 Dutch steamer of 2,229 tons sunk,
Of warships: Minesweeper "Lady Ismay" (495 tons) Dec. 24, 1915.
Also 4 trawlers of around 1,000 tons.

The B-boats carried out outpost and patrol trips off the Thames estuary, the Flemish coast and the Hoofden. "UB10" under Steinbrinck passed the Dover-Calais Narrows and sank two transport steamers off Boulogne on December 20th. (Steamers "Huntly" 1,153 tons and "Belford" 516 tons.)

The Flanders Submarine Flotilla booked this sinking as the first success of a torpedo submarine beyond the Dover-Calais line.

## Mediterranean Sea

Operations by KptLt. Forstmann ("U39"), KptLt. Valentiner ("U38") and one by Rücker ("U34") brought the main result here. He also destroyed a patrol boat when he returned through Otranto Street.

Forstmann sank 5 English, 2 unknown and 1 Greek steamer and 1 barque on the Port Said – Malta steamer route, altogether around 25,000 tons. During the operation Forstmann towed the small boat "UC12", Oblt. Palis, from southern Greece to Port Bardia on the Tripolitan coast. "UC12" was transporting military material there.

---

[109] Footnote in original. Each minefield meant a special undertaking. A total of 180 mines were planted by the 5 boats in December. Weekly stormy conditions severely hampered the activity of the Flanders boats.

Rücker was at sea from December 21, 1915 to January 10, 1916. He first functioned as a war material transport to the Senussi at Port Bardia. When an older small cruiser appeared in front of the harbor when cargo was being unloaded, work was stopped to attack the cruiser. The attack failed. 2 steamers were sunk on the outward journey to Africa and 4 on the return journey. Total tonnage 31,610 tons.

Valentiner was on a long cruise from December 9, 1915 to January 10, 1916. In 1915 he sank 3 steamers totaling 25,297 tons, including the large Japanese steamer "Yasaka Maru"[110] and a large English steamer. He also brought war supplies to Port Bardia and towed "UC12" part of the way there.[111] Jaffa, Beirut and Alexandretta were also visited. Everywhere the boat was enthusiastically received by the Turkish authorities and representatives of Germany and Austria and showered with Christmas gifts. On January 2, war material and another Turkish mission were taken on board in the Bay of Marmaritza. However, the planned delivery of the goods in Port Bardia and landing the mission was not possible due to increased enemy patrol ships. "U38" returned to Cattaro and arrived there on January 10, 1916. Of course, this transport activity and servicing the "Turkish missions" significantly hampered trade warfare. At the suggestion of the KKpt. Kophamel, who had meanwhile become Commander of the Pola Flotilla, the Naval Staff decided to sideline the boats' supply mission activities.

In December, "UB8" (v. Voigt) and UB14", commanded by Oblt. v. Dewitz operated in the Sea of Marmara seeking to destroy enemy submarines.

The overall success in the Mediterranean was around 80,000 tons in December.

---

[110] Typo in original. Corrected here.

[111] Footnote in original. When he lost it while diving under enemy patrols in the Otranto Strait, it was found by "U39" and then finally brought to Bardia by this boat (see above).

Successes in the Mediterranean and Flanders and the plans to renew the trade war that “U24” had been allowed to undertake from the North Sea raised the mood among submariners at the end of 1915 and gave rise to great expectations for success in 1916.

## January 1916

### North Sea

“U20” escorted the steamer "Marie", commanded by the brave captain Sörensen, to sea from Helgoland on January 10 to bring supplies, weapons and ammunition to our East African fighters. Unfortunately, a heavy west storm prevented the escort from being carried out in the intended manner. However, the steamer slipped through the English patrols in the North Sea and later completely accomplished its task.

On January 2, KptLt. Spiegel Frhr. v. Peckelsheim with "U32" returned from a reconnaissance trip that led through Silver Pit to Flamborough Head. The voyage was intended to provide reconnaissance for the minelaying cruisers of the High Seas Fleet. The fleet commander abandoned his intention when Spiegel reported on January 2nd in Wilhelmshaven that the area in question for the minelaying was densely populated by fishing vehicles and covert minelaying was not possible.

"U44" - Wagenführer - started mine operations on January 17, similar to what the boat had done in December 1915, except that Tees Bay and the Blyth area were now the target. The mines were placed near Coquet Island. In pursuit of "U44" two torpedo boats rammed each other, one of which (No. 13) sank on January 26th. A large enemy destroyer escaped only by chance from Wagenführer's torpedo shot on January 24 before Blyth.

In January the I Submarine Half-Flotilla, KptLt. Pasquay, was transferred back to the Commander North Sea Submarines at the same instance as the new boat "U66", KptLt. v. Bothmer.

At the end of January, an enemy submarine ran aground on the sands off Schiermonnikoog Island - apparently on patrol in front of the Ems.

In view of the bad weather prevailing at the time, guarding by outpost boats in front of the Ems was somewhat weakened and we noticed the event too late, namely only when enemy cruisers and destroyers attempted to tow it off the beach. "U32" was dispatched on January 23 to attack these forces and arrived too late. The Englishmen had already started their journey home after abandoning the submarine, taking part of its crew with them. The English boat, it was "H6", was later recovered and brought in by the Dutch.

**Flanders**

The B-boats were heavily used for patrol duty in January; apart from 6 English fishing cutters of 50-70 tons in size, they were therefore unable to achieve success.

The C-boats once again carried out lively minelaying activities which led to various successes.

"UC1" — v. Werner —planted his mineload four times in front of the different outlets of the Thames.

"UC3" — Oblt. Waßner — specialized in Boulogne and Dover and placed his mine load three times in January in front of these ports.

"UC5" — Oblt. Mohrbutter — laid mines once north of Shipwash and twice at Sunk Lightship.

"UC6" —Count v. Schmettow — collided with a darkened fishing trawler at night when intending to lay mines at the South Goodwin lightship and had to return to Zeebrügge for lengthy repair.

"UC7" — Oblt. Haag —, returned from minelaying in front of Stanfort and Hewett Channel on January 1; in January planted a field north of AW Rock, one before the Black Deep entrance and a third at the Humber Estuary.

"UC10" - Oblt. Nitzsche – laid one field at Kentish Knock-Buoy.

The following steamers fell victim to the Flanders mines in January: The English steamer "Traquair" 1,067 tons, sank on January 12 1 nm southwest of Admiralty Pier (Dover). The English steamer "Algerian" 3,837 tons, on the same day $2^{1/2}$ nm southwest of Needles (mines from "UC5", Pustkuchen, placed there in October). Also a "UC" mine sank on January 13, the English fishing trawler "Albion II" 240 tons, at St. Catherines Point.

On the field "UC6" (Count Schmettow) had placed in front of the Humber in December, the English fish steamers "Mediator" (178 tons) and "Courtier" (181 tons) sank on January 2nd and 6th.

Other victims of the Flanders mines in January were 1 French, 3 Dutch and a Norwegian steamer totaling –13,538 tons.

### Mediterranean Sea

In addition to the already mentioned successes of "U34", "U35" achieved some sinkings in January.

Command of this boat was assumed by KptLt. von Arnauld de la Perière on November 18, 1915 (during the overhaul time in Pola) from KKpt. Kophamel. Arnauld had enrolled in the Submarine Service in the summer of 1915 and was ordered to complete his training at Pola Submarine Flotilla. An officer of the highest daring and excellent skill, he was destined to be a great success in the Mediterranean theater. We will see later how often luck astonishingly followed this brilliant commander.[112]

On January 11, Arnauld made his first trip to the western Mediterranean. On January 17, 190 nm southeast of Malta, the English steamer "Sutherland" (3,542 tons) was sunk by gunfire. On the same day, "U35" escaped the performance of the English

[112] Footnote in original. The end of the war by no means concluded Arnauld's fighting. As leader of the assault battalion of the lll. Naval Brigade (von Löwenfeld), he was one of the officers who saved Germany from complete collapse during the inner struggles.

submarine trap "Melanie"[113] due to Arnauld skillfully maneuvered keeping "Melanie's" boat between himself and the trap. The "Melanie" was therefore unable to direct an effective fire against "U35" and Arnauld was able to escape from the enemy fire by diving in time. On January 18th the transport "Marere" (544 tons[114]) was sunk 240 nm east of Malta. An English hospital ship had previously taken the crew of the "Marere" on board. While "Marere" was sinking, 3 submarine hunters carried out an angry artillery fight directe at wreckage drifting "near the steamer", thinking they had the submarine in front of them. Arnauld watched this spectacle amused.

Afterwards the English steamer "Trematon" 4,198 tons was sunk 180 nm east of Malta on January 20th, "U35" had to return to Cattaro because of considerable machinery malfunctions.

Secret orders found on the steamer "Marere" left no doubt that the English defense against submarines was now beginning to be better organized in the Mediterranean, both by arming steamers and by guarding the main trade routes. The need for a 10.5 cm cannon on the submarine instead of the two 8.8 cm began to be felt.

On January 8, Forstmann set out towards Greece with his "U39" heading for the North Aegean Sea under order to attack enemy steamers off Salonika. After much effort he managed to torpedo there the English transport steamer "Norseman" of 9,542 tons on January 22. During the war we had, to the best of our knowledge, recorded this steamer as a sinking success, while it now appears from English White Books that it was actually salved. [115] Here we must pause and provide an explanation. Sending the boats against specific targets was almost always a mistake. The aim of the submarine trade war was to sink as much enemy and contraband tonnage as possible. Set on the fertile trade routes of Port Said-Malta or Malta-Gibraltar Strait, "U39" would have sunk a fine 20-

[113] The correct name is "Margite". Q.ships did sail under a number of aliases.

[114] Typo in original. 6443 tons.

[115] "Norseman" was beached and became a total loss.

30,000 tons with almost mathematical certainty. But in operation, his result was almost zero. From this operational perspective, the reader can easily answer the question of why we did not use our submarines exclusively against the American transports in the second half of 1917 and 1918, but rather distributed them among the trade routes. If we had acted differently, we would have experienced the same many times as Forstmann had with the transport "Norseman" in January 1916.

"U21" - Hersing - set off from Pola on January 22nd. As the success of this cruise falls in February, we will follow it next month.

“UC14”- Oblt. Cäsar Bauer - operating from Cattaro, on January 4th in the morning placed mines at Cap Linduetta in front of Valona. The Italian steamer "Brindisi" of 863 tons sank on January 6th and the Italian auxiliary crusher "Citta di Palermo" of 3,415 tons on January 8th – an immediate success! – On mines "UC14" under Bauer laid in front of Brindisi in the first half of December, the English drifters (minesweeper) "Freuchny" 84 tons and "Morning Star" 97 tons struck on January 8, 1916.

In the Black Sea Oblt. v. Dewitz worked with "UC15" from January 21 to 25, 1916 off the Sulina estuary and the Romanian coast. There were no incidents and on January 25 he returned to Varna.

Officially, the sinking results for January were: 28,477 tons for submarine torpedo successes and 26,022 tons for submarine mine successes. It should be borne in mind that North Sea submarines were almost entirely idle and the Flanders B-boats had only very limited authorization to fire. From a general operational point of view, it was a certain consolation the auxiliary cruiser "Möwe" was active against enemy trade at the time. Of English vessels alone, he sank 27,888 tons in January. The introduction explains why the large North Sea submarines were not allowed to work in the Atlantic according to the Prize Orders unlike "Möwe". This possibility had also been abandoned out of excessive concern regarding American involvement.

## February 1916

## North Sea

As mentioned in the introduction, after the February 11th order was issued regarding treating armed merchant ships as if they were equal to warships; February 24 brought an order to reopen trade warfare on the west coast of England, initially with boats stationed off the Irish south coast. Remarkably within 3 years 1915, 1916, 1917 it was always February that brought decisive changes for conducting submarine warfare. This is not an accident; in February, with the coming of spring, and consequently, good weather every time revived hope of again put to use the new weapon.

In the North Sea, February generally still showed its old character: numerous boats building in the shipyards, ending major winter refits, outpost service, joint work with the surface fleet and escorting auxiliary cruisers. The latter task was carried out by KptLt. Wünsche, who had arrived at the Ems on February 9 with his new boat “U70”. "U70" was at sea for this task from February 25th to March 2nd and did not find any reason to warn the auxiliary cruisers during its training work up cruise to the Norwegian coast.[116]

In February the first of the mine boats, which were under construction at the beginning of 1915 (see Volume 1), was ready for the front.[117]The Naval Staff elected to send them to the Mediterranean. From February 16 to 24, they carried out training and shakedown cruises in the North Sea and to the Norwegian south coast.

The activity at the submarine bases was considerable after the long winter spent refitting the boats. Numerous technical innovations have been tested; Rapid diving mechanisms, which the older boats now also received; Devices to reduce the surge of water when

---

[116] Footnote in original. Auxiliary cruiser "Greif" succumbed in combat with the English auxiliary cruiser" Alcantara "and several destroyers (February 29, 1916).
[117] The large minelayers “U71-80”.

torpedoes were launched from their tubes; develop and introduce new submarine pyrotechnic identification signals that create a smoke column on the surface; experiments with new incendiary bombs to fire wooden hulls; trials with launching mines from submarine torpedo tubes. Since enemy submarines once again undertook numerous attacks against our new submarines in front of the Ems, the outpost flotilla there was reorganized and considerably strengthened. The efforts of the Commander of Submarines to procure modern destroyers for the submarine flotillas as leader boats were unfortunately unsuccessful. This shortcoming, we continued to feel every instance when submarines needed to cooperate with the High Seas Fleet. We shall have to come back to this point in the narratives regarding the fourth phase of the submarine war, in which North Sea submarines were initially used only for military purposes.

On February 24, the Commander of North Sea submarines, FKpt. Bauer, himself gave orders to the submarine commanders in Emden for the new trade war, which, as mentioned above, was little different from trade war according to Prize Rules. All submarine commanders were given new hope that the submarine weapon would now be used effectively via this restricted form; unfortunately, that the new approach would collapse again after just two months was unknown at that time. At the end of the month, "U22" and "U32" went to sea to the English West Coast.

### Flanders

The hopes that had been placed on the arrival of the BII-type boats were finally realized. ("UB18" to "UB47" under construction since April 1915; they had at least two 140-horsepower diesel engines against the one 60-horse engine of the B I boats, which gave them a surfaced speed of around 9 knots. They also had two bow torpedo tubes with four 50 cm torpedoes, a significant increase in combat strength against the two 45 cm torpedoes of the B I boats). (Fig 15.)

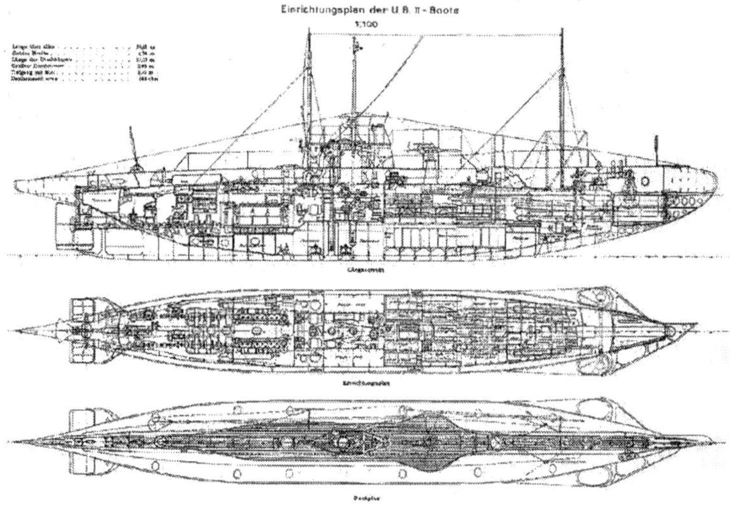


Fig 15. Type BII

The first of these boats, "UB18", arrived in Flanders on February 16 under Steinbrinck's proven leadership; on the 24th he went to Le Havre to disrupt the transport traffic, where on February 26 and 27 he sank the French auxiliary warship “Au Revoir” (1,058 tons) and another auxiliary ship of around 2000 tons. This was followed by the successes of the small B-boats on February 1: A patrol steamer of 700 tons at Kentish Knock ("UB17", Wenninger), a Belgian motor fishing boat and two English fishing cutters by "UB12", Oblt. Keel on February 21; the English steamer "Arbonne" (672 tons) from "UB2", Oblt. Werner Fürbringer.[118]

The activity and successes of the Flanders mine boats were again robust, which gave the enemy reason to take measures against the C-boats operating off the Thames estuary. On February 20, "UC1" was chased by no less than 40 submarine hunters and a destroyer with a search device while attempting to lay its mines off Elbow Buoy. For this reason, Werner had to abandon his original plans

[118] February 26, 1916.

and place his minefield between Tongue Lightship and Elbow Buoy.[119]

Oblt. Wassner ("UC3") laid his mines at the north end of Bassure de Bas on February 22nd. Oblt. Moecke with "UC4" planted his mines at the tip of the ridge of Aldborough Capes on February 10th the English steamer "Cedarwood" (654 tons) and the English steamer "Tergestea" (4,308 tons) fell victim on February 13. Moecke planted another field on the steamer route between Goodwin and the English coast on February 27 in the morning. Minelaying was often made very difficult and disrupted by bad weather.

Mohrbutter ("UC5") planted his mines on the evening of February 9 south of the Knock Lightship and on the evening of February 20 off the Galloper Lightship. On February 21, the English steamer "Dingle" (580 tons) sank there, on February 24, the English steamer "Tummel" (531 tons). The Dutch steamer "La Flandre" (2,018 tons), which sank there on February 21, is probably due to this operation. On the morning of February 27, Mohrbutter placed a field off Calais. Mines laid on three trips by "UC6" (Graf Schmettow) in the Dover and Folkestone area on February 10th, the night of February 20th to February 21st and in the early morning of February 27th brought great success. There sank the English steamer "Leicester" (1,001 tons) on February 12th, the trawler "Carlton" (266 tons) on February 21st (note the tables), on February 28th the trawlers “Angelus” (304 tons) and “Weigelia” (262 tons). The latter two were evidently sweeping for mines in the area on which the English steamers "Maloja" (12,431 tons) and "Empress of Fort William" (2,181 tons) had struck; this clearly demonstrates how Graf von Schmettow, a mine laying specialist, had progressed under difficult conditions off the enemy coast. Oblt. Haag with "UC7" laid mines in the approaches to Harwich early on 10 February[120]; and south of the Galloper Lightship on the evening of February 23. The Dutch

119 Believe there is a typo in the original text as the latter minefield is dated February 9 vs. earlier referenced initial mining attempt dated February 20.

120 Gayer's 1930 edition has added a footnote crediting “UC7's” mines with sinking Commodore Tyrwhitt's cruiser flagship “Arethusa” off Harwich on Feb. 10 despite efforts to tow the ship.

steamer "Mecklenburg" of 2,885 tons sank there on February 27th. Oblt. Nitzsche with "UC10" on the evening of the 2nd planted mines at the Cross Sand Lightship, and on the 22nd in the steamer route near Southwold, where on the 25th the English steamer "Southford" (963 tons) sank, and on February 26th at Kentish Knock.

Successes attributed to the activity of the C-boats resulted in 28,384 tons. In addition, 8 enemy and 2 neutral steamers totaling 17,856 tons sank on Flanders UC mines.

This shows a total result of 46,240 tons for the 7 small Flanders mine boats in February, an operational success, which is to be rated even higher as a number of operations had to be canceled due to bad weather.

### Mediterranean Sea

We saw "U21" after a lengthy repair period depart Pola on January 2 to conduct trade warfare in the eastern Mediterranean. On February 1, Hersing sank the English steamer "Belle of France" (3,876 tons) 126 nm NW-W from Alexandria. Hersing did not keep to the commercial steamer routes, he sensed French warships, and his instinct, unsurpassed in this regard, did not deceive him. On February 8, he sighted the French armored cruiser "Almiral Charner" and sank it with a single torpedo as he had "Pathfinder", "Triumph" and "Majestic". The Operational Command did not entirely agree with Hersing's deviation from the main goal of submarine warfare, working on the commercial steamer routes.

On February 5, Max Valentiner with "U38" first went from Cattaro to Port Bardia in Africa to transport personnel and material for the Senussi. After completing this task, he initially operated on the Malta-Gibraltar line, but was unable to achieve any success there due to the persistently heavy weather. He therefore decided to advance to the Gulf of Lyon, where he was lucky; from the 23rd to the 27th of February he sank 5 steamers and 2 sailing vessels, to which he added the steamer "Springwell" (February 9th, 5,593 tons) and the steamer "Kilbride" (March 1st, 3,712 tons). The total result was 22,752 tons.

Arnauld de la Perière with "U35" cruised from February 19 to March 6 in the sea area between Malta and the Cerigo Strait. On February 23, at 6:20 pm, a large 4-funnel steamer ("Mauretania"?, "Aquitania"?, "Britannic"?) Just escaped his attack as it was already too dark and the shooting distance was too great.[121] He also achieved the following sinkings: February 26th the French auxiliary cruiser "LaProvence" (13,753 tons), on 27th and 28th February each a cargo steamer and on 29th February the English escort sloop "Primula" (1,250 tons). The latter demonstrated an extraordinarily tough life; first its foredeck was destroyed by a torpedo; the crew brought out sailcloth to form a patch, so the ship, steering stern first, could stay afloat. In this condition it even avoided 2 additional torpedo shots.

All respect to the enemy for such seaman-like performance! It only succumbed to another torpedo hit (the fourth shot of the lot).[122] The total result for von Arnauld's cruise was 22,586 tons.

## March 1916

## North Sea

In connection with the renewed trade war activity, "U22" and "U32" departed from the Ems at the end of February. "U22" had to return without any successes due to an accident at the forward hydroplanes and the electric motors, "U32" sank southwest of Fastnet to Bishop-Rock the steamers "Teutonian" (4,824 tons) and Rothesay (2,007 tons), and a French sailing vessel of 250 tons and the French four-masted barque "Ville de Havre" (3,109 tons).

In March, North Sea submarines assigned to trade warfare were "U43", KptLt. Jürst, "U70", KptLt. Wünsche, "U44", KptLt. Wagenführer, and "U28", KptLt. Frhr. v. Forstner. "U43" sailed on March 9 from the Ems to North Channel, "U70" departed on March 10 from Wilhelmshaven to the English west coast. Because of the

[121] A misobservation as all three of these ships were in English waters. There were no other four funnel non-German flagged steamers in existence.

[122] de la Perière referred to HMS Primula as "a tiny wasp" expressing his aggravation regarding expending so many valuable torpedoes at a target of limited value.

early spring, there was considerable bad weather in the north and west of England, which put a lot of strain on the boats; nevertheless, success was achieved. Despite this "U70" sank: 3 English and 1 Norwegian steamer, one Norwegian and one French barque, totaling 19,788 tons which is good for a new boat under unfavorable weather conditions and with very limited freedom of action. The steamer "Berwindvale" (5,242 tons), who was torpedoed by Wünsche on March 16, narrowly escaped sinking. On March 24, a 5,000 ton steamer was also confirmed sunk, the name of which has not yet been established.[123] On March 25, the steamer "Duendes" (4,602 tons) was badly damaged by gunfire, but was still able to reach the port.

"U43", which had the operationally thankless task in front of Ireland's North Channel succeeded in sinking the English steamer "Aranmore" (1,050 tons) 24 nm ENE from Eagle Island on March 21 and on March 24 sinking the English steamer "Englishman" (5,257 tons) 30 nm NE from Malin Head. It was Jürst's first trip to the west coast with a new boat that was technically not yet up to standards.

On March 16, "U68" KptLt. Güntzeh Otto Ludwig departed from the Ems to conduct trade war on the west coast. The boat's successes are not known; it was probably destroyed by the submarine trap "Farnborough" near the Scilly Isles.[124]

Frhr. v. Forstner with "U28" set sail from Helgoland to the west coast on March 18th. The boat's sinking successes in the period from March 26th to April 1st are known: 3 enemy and 1 neutral steamer, 1 English and 1 French sailing vessel totaling 17,694 tons.

On March 27th by Frhr. v. Forstner torpedoed a steamer whose identity could not be determined from the lists.[125]

---

[123] The steamer "Fenay Bridge".

[124] Indeed it was on March 16.

[125] Two British steamers were torpedoed by "U28" on this date.

On March 19, Wagenführ went on his first trip to the west coast with "U44". That day he left the Ems mouth in a thick fog together with "U45", KptLt. Sittenfeld, and "U69", KptLt. Wilhelm. (The latter two newly launched boats went on shorter shakedown cruises to the North Sea.) Wagenführer's first trip to the west coast was to be a great success. He achieved a sinking result of 29,439 tons, a number which would have been enormously larger if Wagenführ had not allowed the large passenger steamer "Mauretania" to pass undisturbed in front of him off North Channel due to the restrictive regulations. The steamer went passed within a few 100 meters of Wagenführ's cross hairs, 4 bow torpedoes were ready in their tubes, the tube doors were open, the commander's index finger was on the push button of the electric firer; the shooter was assured of a score over 30,000 tons. Wagenführer resisted, obeyed orders; today we know from an English sources that our enemies regard unrestricted warfare against enemy merchant ships by submarines as a legitimate means of "naval warfare" - as a struggle against enemy lines of communication - let us learn from this for the future!

While these boats were operating far from home, Admiral Scheer tried to get the new boats, which were to be used for North Sea operations, to work with the fleet. The new mine boat "U74" was supposed to plant its mines in front of the Firth of Forth in connection with a fleet sortie. Unfavorable weather prevented the fleet's plans. "U74" alone laid its mines on March 31 on the May Island - St. Abbs Head line. The English steamer "Sabbia" (2,802 tons) ran into this field on April 20.

Towards the middle of the month, the enemy countermined the area west of the Ems estuary. The blockade breaker "Ottensen" ran onto mines the night of March 16, March 17, 17 sea miles WNW of the Wester-Ems control barrel buoy. Fortunately, however, it soon became apparent that the exit route currently used by submarines remained mine-free via the Dutch islands to Ameland and then course NNW.

On the March 25th the English light forces entered the German Bight with the intention of attacking the airship hangar in Tondern with airplanes. Enemy forces were first sighted at Sylt. The fleet immediately took action against the enemy, "U69" was dispatched north from the Ems, but was unable to find the enemy. In the operation, we lost the torpedo boats "S22" and "G194", the former probably due to the explosion of its own torpedo, the latter due to the ramming by the English cruiser "Cleopatra". On the English side, the destroyer "Medusa" was lost due to a collision with another English destroyer. The English cruiser "Undaunted" collided with "Cleopatra" during the attack on "G194". "Undaunted" was badly damaged, but reached the port.

"U66" under Kptlt. v. Bothmer departed on March 29, 1916 on his first long-distance cruise to the west coast; we shall return to it in the April results.

**Flanders**

The stock of high-performance boats was further strengthened by the addition of "UB19", Kdt. Oblt. Becker, on March 1st and "UB29", commander Oblt. Pustkuchen, on March 8th.

"UB8" and "UB29" went to sea on March 7th; "UB18" to the French coast and "UB19" to the English coast to disrupt transport traffic across the Channel. "UB18", Steinbrinck, sank the English steamer "Harmatris" (6,387 tons) 14 nm NW-N from the Boulogne mole on March 8, the French transport steamer "Lousiane" (5,109 tons) and on March 9 another ship off Le Havre.[126]

"UB29" sortied for the first time north of Outer Gabbard Sand on March 19 to conduct trade warfare and sank the French steamer "Nominoe" (3,155 tons) and 2 other steamers on March 20.

Steinbrinck had further successes on March 23rd at Le Harve roadstead. There the English steamer "Kelvinbank" (4,209 tons) and 2 unknown steamers were sunk. [127]

[126] Norwegian S/V "Silius".

[127] Contemporary records credit Steinbrinck with one other steamer.

On March 29, 1916, soon after Steinbrinck returned from this cruise, the Flanders Submarine Flotilla celebrated its one-year anniversary. Steinbrinck, who had achieved such excellent results with the small, weak "UB10" and "UB18" boats received the well-deserved Order Pour le Merite medal. Steinbrinck's achievements have been outstanding in the history of the submarine warfare.[128] This has been fully recognized by all of his comrades, even the most important, without envy.

In those days occurred the fateful torpedoing of the French steamer "Sussex" (1,351 tons) by "UB29", Oblt. Pustkuchen, on March 24th. When targeted, the submarine commander considered the "Susssex" to be a transport due to the large number of persons visible on deck. Since she was a passenger steamer, and after successful salvage the investigation determined the torpedo origins without any problems, England once again had the opportunity to score a coup against us with America. We will come back to this later. In addition to torpedoing the "Sussex", Pustkuchen succeeded in sinking the English steamer "Salybia" (3,352 tons) 4 nm SW-S from Dungeness on March 21.

Not only had the newer boats "UB18" and "UB29" successes, but also the older small boats. "UB13", Oblt. Neumann, destroyed 4 English fishing trawlers (76 tons each) off Hoofden, "UB17", Oblt. Metz, on March 6, 2 English fishing trawlers at Yarmouth.

“UB6”, Oblt. Voigt sank a large steamer between March 16 and 18 at the North Hinder lightship and on March 31 at the Nord Galloper buoy.

“UB10”, Oblt. Salzwedel, on March 19, destroyed the English tanker "Port Dalhousie" (1,744 tons) 2 nm S1/2W from the Kentish Knock Lightship.

---

[128] Steinbrinck had sunk 11 steamers and several dozen smaller craft as of his medal investiture. He would sink several dozen more steamers before the end of the war, finishing as one of the top commanders in terms of hulls and tonnage sunk.

"UB13", Oblt. Metz, a steamer each on March 16th and 31st.

The compilation shows a pure, historically confirmed torpedo result for Flanders in March of 56,191 tons.

Flanders mine successes were again significant. The following minelaying sorties were carried out:

| Boat | Cdr. Name | Sortie Date | Minefield location | Remarks |
|---|---|---|---|---|
| UC1 | Oblt. v. Werner | March 17 to 18 | North Hinder Lightship | 24 mines |
| | | March 20 to 22 | SouthWest of Kentish Knock | |
| UC3 | Oblt. Wagner | March 6 to 7 | Harwich Approaches | 24 mines |
| | | March 16 to 17 | East of Kentish Knock | |
| UC4 | Oblt. Möcke | March 3 to 5 March 11 to 12 | Off Tongue lightship on the trade routes between the Goodwin Bank and the English coast | 24 mines |
| UC5 | Oblt. Mohrbutter | March 6 to 7 | West Sunk Lightship | |
| | | March 16 to 17 | East of the Sunk Lightship. | 48 mines |
| | | March 24 to 25 | South of Inner and Outer Gabbard Sands | |
| | | March 30 to 31 | South exit of the Hewett Channel At Dover | |
| UC6 | Oblt. Graf v. Schmettow | March 3 to 5 | | 48 mines |

| Boat | Cdr. Name | Sortie Date | Minefield location | Remarks |
|---|---|---|---|---|
| UC6 | Oblt. Graf v. Schmettow | March 9 to 11 | At Dungenes | |
| | | March 16 to 17 | To North of South Foreland | |
| | | March 22 to 23 | Off Folkestone | |
| UC7 | Oblt. Haag | March 8 to 9 | N of North Foreland LightBuoy | |
| | | March 14 to 15 | At the mouth of the Thames | 36 mines |
| | | March 24 to 25 | North of Elbow buoy on the paths leading to Boulogne | |
| UC10 | Oblt. Nitzsche | March 6 to 7 | Northern Exit of Black Deep | |
| | | March 12 to 13 | NW of Longsand Floating Lightship | 36 mines |
| | | March 17 to 18 | N of North Galloper buoy | |
| | | | Total | 240 mines |

As far as it can be determined according to historically confirmed information, 25 mined vessels totaling 28,295 tons fell victim in these minelaying operations.

The total success of the Flanders boats in March 1916 was therefore 84,486 tons.

## Mediterranean Sea

The result in March in the Mediterranean was relatively low, because after their almost uninterrupted activity since arriving in the Mediterranean war theater, the boats there had to visit the shipyard in Pola for a lengthy overhaul.

Of success there were:

Arnauld de la Perière with "U35" was on cruise from March 16 to April 8 in the Eastern Mediterranean. On this trip, however, only the English steamer “Minneapolis” was sunk - 13,543 ton 195 nm W½N from Malta. Several times, good vigilance on the part of steamers hampered success.

Oblt. v. Heimburg operated in the Black Sea with "UB14" from March 5 to 27 to protect the Turkish wing of the Caucasus Army near Atina, which was threatened by Russian naval forces. The boat’s reported presence was enough to keep Russian ships away.

In the same area Gansser operated with "U33" from March 25th to April 7th; on March 30, he sank the Russian transport "Portugal" (5,358 tons). On March 31, Gansser destroyed the Sukhum Lighthouse with artillery and fired on the city of Gradant in the afternoon. On April 3, "U33" escorted the cruiser "Breslau", which fired at the Russian positions at Kannaghiam. A sailing vessel of about 70 tons and a steamer of about 1,500 tons were further prey on March 31st.

Oblt. Dewitz with "UC15" left Varna on March 27th and planted a minefield off Sevastopol on March 24th.

"UC12" under Oblt. Fröhner planted a minefield off Brindisi in the afternoon on March 2, in-between the passage of Pedagne Island and Castello Mare. On March 12, Fröhner left Cattaro for another mining operation off Taranto. The boat would not return from this venture; in the Gulf of Taranto, a mine ran out and sank the boat; the boat was later raised and restored at an Italian shipyard. The war diary found gave the enemy information regarding the boat's previous activity. On March 8, the English steam drifter “Enterprise II” (84 tons) ran on the mines in front of Brindisi.

Oblt. Franz Becker laid mines with "UC14" on March 18 in the afternoon at the Sivota Island (in front of Corfu’s south channel). On March 15th the boat was shot at by an enemy submarine with 2 torpedoes. A torpedo hit the rear edge of the conning tower, but did not detonate due to the failure of the ignition device. The French fishing trawler "Ginette" displacing 272 tons ran onto "UC14's" mines on March 29.

With the steamer "Kilbride" sunk by "U38" (KptLt. Max Valentiner) on March 1st, the result for March in the Mediterranean and Black Sea was 25,539 tons of ship capacity.

The total result by submarine torpedoes and mines was around 190,000 tons in March.

## April 1916

The submarine war, which was intensifying again in March in the North Sea and in Flanders, was on a good start to success, the following text will demonstrate it would have produced desired results; instead the Sussex Note imposed an increasingly stringent warfare.

### North Sea

The following new boats commenced operations. "U66" under KptLt. Bothmer, "UB21" under Oblt. Hashagen, "U69" under KptLt. Wilhelms, "UB22" under Oblt. Putzier, "U45" under KptLt. Sittenfelsd, "UB27" under Oblt. Dieckmann, "U73" under KptLt. Sieß.

Bothmer, as mentioned, left the Ems on March 29, sank 6 enemy and 1 neutral steamers totaling 19,502 tons from April 5 to 10.

An English cruiser of the Argyll class narrowly escaped Bothmer's attack on April 15. Bothmer, as his success demonstrates, commenced war operations with considerable vigor and energy.

"U22", Kptlt. Hoppe, went to North Channel on April 1st. On April 6 he torpedoed the English steamer "Vennacher" (4,700 tons) 28 nm WNW from Skerryvore the steamer could still be brought to Lough Swilly; on April 8, the English steamer "Adamton" (2,304 tons) was sunk 15 nm south of Skerryvore by gunfire. Added to this was the sinking of the English steamer "Chic" (3,037 tons) 45 nm southwest of Fastnet (total 5,341 tons) on April 13. Seeing "U22" results one has to take into account this was a rather unwieldy boat when submerged.

On April 1, "U73" under KptLt. Sieß departed Cuxhaven using the north around England route to the Mediterranean. The boat was equipped with 34 mines. On April 11, Sieß, 108 nm WNW from Fastnet, sank the English bark "Inverlyon" (1,827 tons) by artillery fire. On April 17, in front of Lisbon (Portuguese shipping had been seen as hostile since March 9), mines were laid as planned in front of the north and south entrances. The first field the boat managed to plant while surfaced. On the same day the Norwegian steamer "Terje Viken" (3,579 tons) sank on these mines. On April 19, the mining operation planned before Gibraltar was abandoned because "U73" had already been seen west of the straits and the bright moonlight made the operation difficult. On April 27, shortly after midnight, a field of 22 mines was laid about 1 nm from Valetta (Malta), initially while surfaced, then submerged when destroyers came into view. While departing from the mining site, an English battleship steering zigzag courses to Malta was spotted at dawn. It promptly ran into the middle of the field just laid by Sieß and sank; it turned out to be "Russell" (14,000 tons). On the same day the sloop (war vessel) "Nasturtium" (1,250 tons) ran on the mines and sank as did the armed yacht "Aegusa" (1,242 tons), which was mined on April 28. When Sieß arrived in Cattaro on April 30th, he was able to look back on a great success; his operation is the typical example of the success one can hope for from unexpected operations in naval warfare. The stronger the means that can be used in such cases, the greater the effect. With submarines we unfortunately only managed to execute such unexpected strikes with one boat at a time because in most cases we were unable to send more.

On April 4, "U44", whose successful journey was described in the month of March, reported its location off Scotland with a 600 nm transmission distance. This was a record achievement by the previously mentioned Senior Telegraphist Mate Degen commanded by Oblt. Günther Krause.

On the same day, "U69" under KptLt. Wilhelm departed the English west coast. This excellent submarine commander was previously commander of "UA" in the Baltic Sea (see footnote p.91). He began his operations in the North Sea with great success. Wilhelms only carried out successful cruises until July 1917, when a still unexplained fate ended his activities.

On this cruise he sank 4 steamers and 4 sailing vessels totaling 21,051 tons between April 15 and 20. The sinkings took place in the area south of Ireland to the Scilly Islands.

On April 7th, "U67" and "U20" set sail from Borkum-Reede. FKpt. Bauer, The Commander of Submarines, took passage on board "U67" to personally obtain a view of the conditions in the submarine operations areas on the English west coast. Unfortunately, while "U67" was successful on this trip sinking a steamer and two sailing vessels of 8,989 tons altogether in the sea area south of Fastnet Rock and at Ushant it came across 2 very cleverly commanded traps who severely damaged the boat with artillery fire. It is understandable the Commander of Submarines, FKpt. Bauer, thereby gained an unfavorable opinion regarding the feasibility of conducting submarine warfare where steamers required warning prior to commencing any shooting. Boats' experiences, on average, were not nearly as adverse as FKpt. Bauer found them on "U67" is evident from favorable results achieved by most other boats at this time. We will have to come back to this point.

"U20", still under the leadership of the tried and tested KptLt. Schwieger, suffered a misfortune on April 13, when the forward hydroplane was knocked loose by the swell. On the night of the 13th and 14th Schwieger reported the incident and “U19", which was now commanded by KptLt. Raimund Weisbach, left to the accident site and took its place. ("U19’s" engines had been repaired and modernized in the three-month refit period at the Imperial Works in Wilhelmshaven, so the boat could now be sent to the English west coast.) On the morning of the 15th Weisbach left the Ems and went north the same day. Weisbach showed great skill in this his first cruise. He smoothly executed a special mission[129], he also sank 5 enemy steamers and 1 Norwegian sailing vessel totaling 19,280 tons. On May 1st, Weisbach returned to Emden after this prompt completion. The old "U19" had proven itself excellently. The conversion of the engines was a complete success.

Last of the new large boats, "U45" under KptLt. Sittenfeld departed in April for its area of operations off southern Ireland. Sittenfeld succeeded in sinking two steamers and two sailing ships totaling 8,031 tons in the sea area west of Ushant. On April 25 the Fleet Commander issued an order: "All submarines including those on west coast are to return home; trade warfare only to take place according to Prize Rules." Sittenfeld had not received this message due to the great distance. From April 25th, no more submarines departed to conduct trade warfare in the North Sea. Commander of Submarines, FKpt. Bauer, was absent from Wilhelmshaven in the days when this decision was made, which we will look in more detail how it evolved at Wilhelmshaven. It was only on April 30 that he entered Wilhelmshaven with "U67".

"U20" (Schwieger), who had departed the Ems early on April 24, did not receive the signal to return, and thus became the boat that brought this trade war period in the west of England to a close. Three steamers and 3 sailing vessel a totaling 23,162 tons (including "Cymric"=13,370 tons) were his quarry.[130] On May 1, "U20" reached a record radio range of 770 nm by sending a location and weather signal from the Bay of Biscay. This was an excellent performance by Senior F.T. Heinke from the Staff of the III Submarine Flotilla. During this trip, the boat did not sink various neutral vessels and even a Russian one because the crews could not be saved with certainty. This should be noted by everyone who sees articles in the enemy press regarding the "Barbarism of German submarines".

---

[129] The special mission was to land Sir Roger Casement and two agents in Ireland ro foment an uprising against British rule.

[130] Footnote in original. These results indicate the extent to which the operation could develop if it was decided to conduct unrestricted submarine warfare; 20,000 tons was considered the average performance for each large submarine operating on the English West Coast.

Towards the end of April, the new boats "UB21", "UB22" and "UB27" operated in the North Sea. "UB27" under Oblt. Dieckmann sank neutral and English vessels totaling 5,964 tons during a cruise from April 22 to May 5 in the North Sea. Dieckmann was subsequently a great success; He became a master, especially when working close to the enemy coast and in narrow, guarded traffic routes; he went missing from a daring trip to the Irish Sea in 1918.

Admiral Scheer kept the fleet in motion during this trade war activity. Scheer's efforts to strike at the enemy was unmistakable; intended mine laying by "U71", KptLt. Hugo Schmidt, in the Firth of Moray and "U72", KptLt. Krafft, in the Firth of Forth were to support fleet movements. Unfortunately, on April 21, both boats returned unsuccessful; Due to engine failures and oil bunker leakage respectively, and they had not been able to lay their mines where instructed. The Fleet's sortie occurred on the 24th. The scope of this text only allows for a brief summary. April 25, our battle cruisers fired on Lowestoft and Yarmouth. Several North Sea submarines were sent to the border area of the German Bight to support the returning fleet. The Flanders Submarine Flotilla supported the operation by positioning boats with red lights displayed to aid in approaching the enemy coast at Lowestoft.

From a technical point of view, it should be mentioned the older boats at the front were gradually equipped with newly installed quick-diving devices. These installations required much testing with the boats practicing off Borkum-Reede and Helgoland. In addition, there were experiments with underwater signaling to establish a perfect connection between surface vessels and submarine or between submarines. Various advances were made at this time. In mid-April, "U46" achieved a mutual connection with ship and land stations at 120 nm range on a training trip to the North Sea using the wave deflector now set up as a radio antenna.

The activity of the North Sea submarines in April resulted in a sinking total of 86,455 tons, which would have been even greater had the cessation order not been issued on April 24, when everything was proceeding nicely.

## Flanders

Regarding torpedo successes, April was again characterized by the brilliant achievements of Oblts. Steinbrinck and Pustkuchen ("UB18" and "UB29"); who were joined by some successful small B-boat cruises. “UB26”, Oblt. Smiths, on April 5 got into a submarine net at Le Havre, from which he could not escape; the boat had to be blown up by the crew; who were captured by the French. "UC5", Oblt. Mohrbutter, left Zeebrugge for a mining operation on April 26, it was known on April 27 from enemy wireless signals that the boat had apparently gotten stuck at Shipwash Sand and had been blown up by the crew; in this case too, the entire crew was captured. "UB26" was later recovered by the French, "UC5" by the English.

This month a new enemy measure began to complicate the work of the Flanders boats. In the second half of April, the Flanders boats discovered net barrages in the area of Thornton-Bank and thence parallel to the coast, into which submarine mines were woven. “UB10”, Oblt. Nitzsche got into this network on the 24th and took 8 hours to escape from it; there were numerous detonations near the boat, apparently from the net mines. The boat had to be dry docked after returning to base because there was still nets with mines attached wrapped around the screws. At the time it was still unclear whether these nets were only intended as a defense against submarines leaving Flanders or at the same time were preparations for an attack on the Flanders coast; later it turned out that it was the beginning of a complex anti-submarine defense. .

In April, 16 vessels of 23,324 ton were sunk by torpedoes from the Flanders boats, including the English submarine "E22", which Steinbrinck sank at Braune-Bank on 25 April.

For the April 25 Fleet sortie, Flanders dispatched “UB10”, “UB12”, “UB13”, “UB18” and “UB29”, “UC6” and “UC7” against the English coast. Except for the destruction of “E22” by “UB18”, the boats were unable to achieve any success, since our fleet did come into contact with the enemy to any significant extent.

The following overview shows the mine sorties made by Flanders in April:

| Boat | Cdr. Name | Sortie Date | Minefield location | Remarks |
|---|---|---|---|---|
| UC1 | Oblt. Werner | April 4 to 5 | Shipwash at Orfordness | |
| | | April 10 to 11 | South of Kentish Knock | Strong opposition from enemy patrols 5x12 = 60 mines |
| | Oblt. Hansen | April 20 to 21 | at Galloper North buoy | |
| | | April 24 to 26 | East of Kentish Knock | |
| | | April 28 to 30 | at Galloper North buoy | |
| UC5 | Oblt. Mohrbutter | April 9 to 10 | North of North Galloper buoy | Attacked by a airplane (two bombs). |
| | | April 21 to 22 | at Sunk LightShip | 2x 1 2 = 24 mines |
| UC5 | Oblt. Mohrbutter | April 24 | to Shipwash | sortie aborted due to enemy activity |
| | | April 24 | to Shipwash | April 26 ran aground on sandbank. Boat blown up. Later recovered by enemy. |

October 1915 to the end of April 1916

| Boat | Cdr. Name | Sortie Date | Minefield location | Remarks |
|---|---|---|---|---|
| UC6 | Oblt. Graf v. Schmettow | April 3 to 4 | To Calais | |
| | | April 6 to 7 | West of Folkestone Gate | |
| | | April 20 to 21 | at Dyck Lightship | In the fairway to Dunkirk |
| | | April 24 to 25 | At the north entrance to Shipwash on the Orfordness — Shipwash lighthouse line. | Counterattacked by destroyers 4 x 12 = 48 mines |
| UC7 | Oblt. Haag | April 1 to 2 | a Sunk Lightship | |
| | | April 8 to 9 | NW and SW of Tongue Lightship | April 24 off Flanders coast. Numerous minesweepers and destroyers, coast alarmed |
| | | April 20 to 21 | South of Thames Estuary | April 27 attacked by a French airplane dropping bombs. |
| | | April 24 to 27 | West of Shipwash and near Northern tip of Bawdsey buoy | 4x12 = 48 mines |
| UC10 | Oblt. Nitzsche | April 1 to 3 | SW of Kentish Knock | |
| | | April 7 to 8 | South of South Inner Gabbard Buoy | |
| | | April 21 to 22 | South of Longsand Lightship | |
| | | April 28 to 30 | In steamer route near Southwold | 4x12 = 48 mines |
| | | | Total | 228 mines |

In April 1916, 18 vessels totaling 23,662 tons sank on Flanders mines, including 4 naval trawlers.

The Flanders April result thus amounted to 46,686 tons, a significant decrease compared to the 81,486 tons in March, although the April 25 return to base order had a significant impact. Increased defensive countermeasures, consistently developed by the enemy, played a role that should not be underestimated, both on the steamer routes and on C-boat approach routes.

In fact, the heyday of the Flanders submarine mine success was already over; this activity was only somewhat refreshed by the intervention of the CII boats (enlarged mine boats) at the end of 1916; but even then, operating conditions were not as favorable as in the first months of the Flanders C-boat mine offensive. (See Fig 16 for an example.)

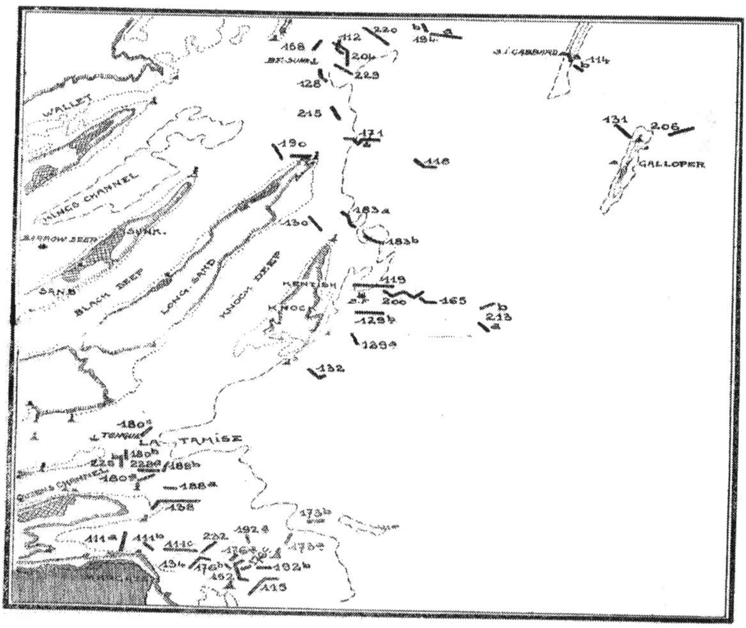


Fig 16. Mining in the Thames April to October 1916. Dark bars represent the fields. Numbers are German reference identifiers.

## Mediterranean Sea

In April the Mediterranean was marked by two successful cruises by KptLt. Rücker ("U34") and KptLt. Forstmann (Walter, "U39").

Rücker operated in the western Mediterranean from March 27 to April 18. On March 29, when he was passing Otranto Street, he had to deal with 4 enemy armed trawlers (here, too, enemy patrols gradually became more consistent). By April 18, the day he returned to Cattaro, he sank 7 English, 1 French steamer and 1 Russian sailing vessel of a total of 28,707 tons.

Forstmann left Pola on March 29th, initially seeking the lost "UC12" in the Gulf of Taranto, then went between Tunis and Malta, leaving strong evidence of fine activity there, then headed for Cape Spartivento (Sardinia) and now went to the Gulf of Genoa, especially before Toulon. The way back led south of Sicily through the Otranto Strait to Cattaro, where Forstmann arrived on April 15th. 5 enemy, 2 neutral steamers and 2 enemy sailing vessels totaling 19,850 tons were sunk.

"U21" (Hersing) set sail from Brioni (near Pola) on April 10 to transport a mission of German army officers and Turks to the North African coast. From April 15th to April 20th, three attempts were made at places in Greater Sirte (Gahndie, Zuitana and Marsa Brega) to hand over the transport. Only on April 16 in the third location was a reasonably safe berth and land connection possible. The transported arms and ammunition was unloaded on the 19th. The Turkish and the German army officers could not be deposited due to the uncertain and unreliable conditions on land so they had to be taken back to Pola. On April 30, Hersing sank the English steamer "City of Lucknow" (3,677 tons) 60 nm east of Malta; on May 4th, "U21" returned to Cattaro. The activity of this boat had been adversely affected by the transport and numerous machinery malfunctions.[131] "UC14" (Oblt. Franz Becker) operating from Cattaro in April, made two mine trips, on April 6 on the steamer route east of Bari and April 16 on the steamer route between the island of Corfu and the island of Naxos. The French steamer "Jean Bart" (475 tons) probably sank on the mines of "UC14".

"U33" (Gansser), who was absent from the Mediterranean, departed Constantinople to the eastern Black Sea from April 18 to May 4 to prevent Russian operations against Trebizond. Apart from various battles with land positions, where Gansser fortunately prevailed, on April 24, he sank the Russian steamer "Kiev" (3,000 tons) and three sailing vessels; on May 4, he entered the Bosphorus again.

"UB7" (Oblt. Werner) sank 2 sailing vessels on a voyage from Varna, which was used to explore Russian shipping traffic between Sevastopol and Odessa-Nikolaev (April 2nd to 11th, 1916); under a new commandant (Oblt. Lütjohann, the boat undertook another trip from April 23 to May 4 to the south Crimean coast and operate against enemy warships. However, she only had an artillery engagement with a sailing vessel and returned to Varna on May 4.

"UC15" (Oblt. von Dewitz), also operating from Varna, placed a minefield off Sevastopol on April 1st and 30th, and on April 25th one in the steamer path northeast of Cape Chersonese. A sailing vessel was also sunk on this trip.[132]

## Review of the 3rd Submarine Warfare Period.

If we were to describe the events in the Mediterranean and Black Seas on a monthly basis, one would not see any fundamental difference between future periods of warfare compared with the previous. This stems from Chief of Mediterranean Submarine Operations, KKpt. Kophamel, describing as perfectly acceptable the Admiralty Staff's demand for trade war only according to Prize

[131] Footnote in original. It was decided to refuse such transports in the future if their complete execution, i.e. Quick disembarkation from the transports did not seem guaranteed.

[132] Footnote in original. Total tonnage sunk in the Mediterranean and the Black Sea in April was about 55,980 tons.

Rules, even if an unrestricted submarine war promised greater success. The years that followed proved KKpt. Kophamel correct. The prohibition on torpedoing ships without warning did not reduce activity in the Mediterranean; on the contrary, the success rate there increased monthly and in August 1916, achieved around 140,000 tons, almost the monthly average of what was later sunk with a significantly larger number of boats during the unrestricted submarine war in the Mediterranean.

The ultimatum presented by Fleet Commander Admiral Scheer to the Naval Staff, to either intensify submarine warfare or only employ the boats in purely military operations, but in no case to continue trade warfare according to Prize Rules, was initially approved at home by all war-experienced submarine commanders, since it was seen as a means to eliminate the constant wavering of the High Command.

The Commander of the Flanders Submarine Flotilla, KKpt. Bartenbach, agreed with the opinion of the Fleet Commander, KptLt. Schwieger (commander of "U20"), who, as a particularly experienced submarine commander, was asked to report, expressed considerable pessimism regarding the possibilities of waging submarine trade warfare according to Prize Rules. So long as one could hope to rescind the April 25 order ("to only conduct trade warfare according to Prize Rules") and instead continue the previous policy (authorizing destruction without warning of enemy merchant ships in the declared war zone and troop transports and armed merchant vessels found anywhere), it was agreed to reject trade warfare according to Prize Rules. But when it became clear that we folded before Wilson's "ultimatum note" of April 20[133], the majority of the North Sea submarine commanders and soon also the Commander of the Flanders Submarine Flotilla changed their opinion. Two years of practical war experience had shown that strictly "military use" meant the nearly 30 large boats would largely

133 Footnote in original. See Tirpitz page 367. [Alfred von Tirpitz, "Erinnerungen" Leipzig: K.F.Koehler, 1919.]

lay idle; because meager results – except for those relating to supporting the High Seas Fleet going into battle – came out of "purely military use" of the boats. Flotilla commanders and Ems submarine commanders[134] were therefore keen to meet with the Commander of Submarines and Fleet Commander in Wilhelmshaven and change this decision. It was stated, particularly in the coming summer months, trade warfare according to the prize rules outside from guarded enemy areas was promising and the danger to boats minimized by means of suitable procedures. (Stopping steamers at long distances, approaching stopped steamers submerged, and possibly using a captured vessel for steamer examinations.) These proposals did not fail to impress in Wilhelmshaven, but the Commander of Submarines was missing in these crucial days (he was still at sea with "U67") and when he returned, the Fleet Commander had already set himself too much in opposition towards the Naval Staff, moreover FKpt. Bauer was currently pessimistic due to his unfavorable experiences on "U67" (see page 149) regarding the possibilities of trade war according to Prize Rules, the report of the KptLt. Schwieger did the rest.[135] Trade warfare according to Prize Rules was rejected against the advice of experienced commanders and submarine leaders, who were known to be cautious rather than reckless when assessing operational options and who particularly pointed out the resulting decrease in tonnage sunk could not be compensated for later. It was the conviction of those who spoke in favor of trade warfare according to Prize Rules that it would not be permanent and after a few months there would be introduced relief for the submarine commanders, i.e. a loosening of restrictions. Despite rejection of the Fleet Commander by the Naval Staff (Admiralstab) trade warfare according to Prize Rules could have been ordered but was not issued and did not occur in May, June, July, August and September

[134] Footnote in original. At that time almost all war-experienced boats were stationed there.

[135] Footnote in original. It was particularly unfortunate Schwieger had already departed on April 24, as he could not be called to the meetings. Otherwise he would have had sufficient self-awareness to recommend waging trade war according to Prize Rules rather than leave the boats unused.

1916 for North Sea submarines.[136] In Flanders, KKpt. Bartenbach started trade warfare according to his own responsibility at the beginning of September because he could no longer stand watching his boats lay idle. Tonnage not sunk during this period due to abandoning submarine trade warfare, according to careful estimates, is approximately 1,200,000 tons for North Sea and 400,000 tons for Flanders boats. It was only on October 6, 1916, the Naval Staff issued orders to commence general trade warfare according to Prize Rules. What opportunities were possible is illustrated by the new boat "U49" which sank over 40,000 tons on its maiden voyage to the Bay of Biscay in November 1916. Some have argued the failure to sink 1,600,000 tons due to the non-conduct of trade warfare according to the Prize Rules had a decisive influence on the fate of the submarine war. One has to agree with this judgment since the end of the war even our enemy has sometimes described the suspension of the trade war in 1916 as England's salvation.[137]

In any case, England's salvation lay not only in the interruption of the trade war but also to a large extent, fluctuating operational policies, such as the period from October 1915 to the end of April 1916 and the period that followed, the impetus for an extensive submarine building program was completely absent. It would have taken a rather brilliant, foresight among those responsible for submarine construction to realize, after all the trials and tribulations, that an intensified submarine war would arise in February 1917 and accordingly design a generous construction

[136] The 1930 edition of this work has additional text in this section indicating after his return, FKpt. Bauer submitted several recommendations regarding continuing trade warfare to Admiral Scheer who passed them onto the Naval Staff. Admiral Scheer's Chief of Operations, Kapitän von Levetzow returned from Berlin at this time reporting the political chiefs and Naval Staff's extremely unfavorable opinion regarding escalating at any time the submarine war and in particular an unrestricted one. This news strengthened Scheer's resolve for the only manner to obtain authorization to wage unrestricted submarine warfare was to veto any other method.

[137] Footnote in original. The "1.6 million tons lost" estimate is the lowest of all available estimates. The submarine arm was said to have missed sinking 2 to 3 million tons during the war.

program in March 1916 when attempts were being made to commence a more vigorous war on trade. This mindset was lacking, during the period described and as a result only the final 31 CII boats, whose construction was planned since Spring 1915 (volume 1), were finally ordered in January 1916. The next construction contract, which was by no means a "large program", took place in May 1916. (22 large torpedo submarines, 24 small torpedo submarines, 10 large boats. See Figs. 17 & 18.)

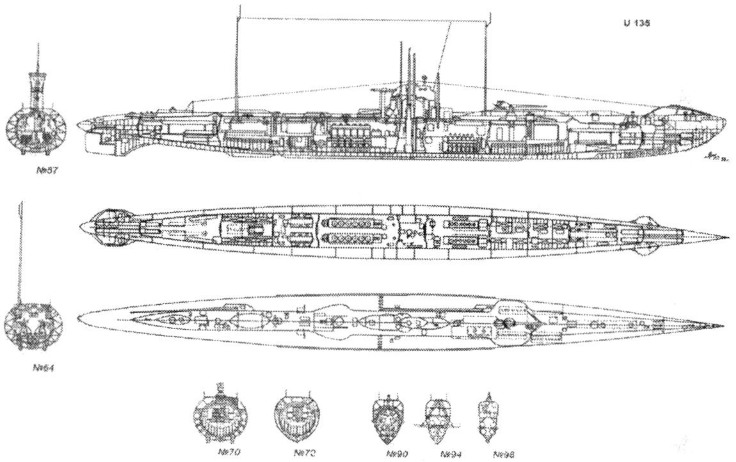


Fig 17. U 135 large type submarines ordered May 1916.

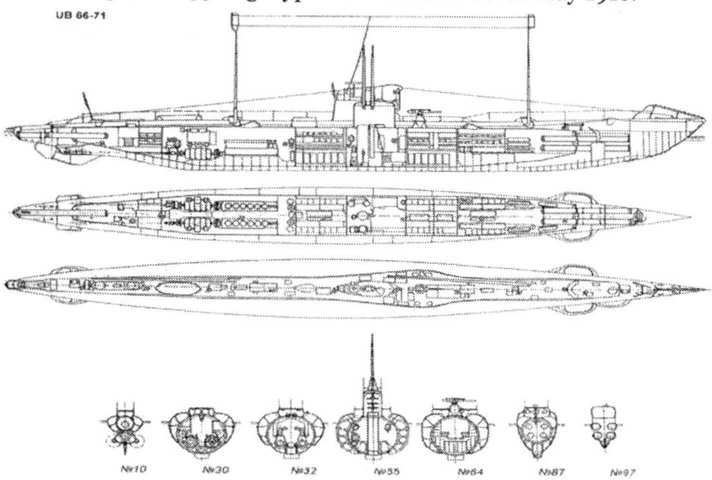


Fig 18. UBIII type small submarines ordered May 1916.

It remains to brief provide an overview of the boats that were faithfully performing their duties in the Baltic Sea, far removed from the main submarine theaters.

Boats "U1" to "U4" were now finally pulled out of the war and sent to serve at the Submarine School.

"UA"[138] was under Oblt. Scherb worked until April 1916 from Libau. These cruisers were mostly used to search for enemy submarines, to observe the western entrance to the Gulf of Riga (February 1916) and to support our own armed forces laying minefields in the Gulf of Finland (March and April 1916).

"UC4" (Oblt. Vesper) placed a minefield at Vidskär and Utö on November 2, 1915, then the boat was transferred to the Flanders Submarine Flotilla, where its experiencees (under the leadership of Oblts. Moecke) have already been described.

"U9" (Oblt. Spieß) made one cruise each in October, November, December 1915 in the Gulf of Finland.

On November 5, Spieß carried out a torpedo attack on a Russian minelaying formation off Odensholm; a minelayer was sunk. In March and April the boat went on training cruises in the North Sea, and then it finally entered the Submarine School.

"U10", KptLt. Stuhr was in the Aland Sea from October 30 to November 8 to determine where enemy submarines came from. On November 6, a Finnish steamer was sunk. From December 30, 1915 to January 3, 1916, a special undertaking was carried out to guard the Gulf of Riga at Zerel. On May 27, 1916, Stuhr set out to a holding position NW of Gotska Sandö. “U10” never returned from this venture.

"U17" under KptLt. Hans Walter undertook a cruise from October 19 to 20, 1915, to the Aland Sea to intercept the Swedish steamer "Wally" and guard the area between Bogfkär and Svendskär

[138] Typo error in original which read “U9” Corrected in text above.

Högarne; 2 steamers were brought to Libau. From November 27, 1915 to December 8, 1915 a special venture to the Aland Sea to stalk enemy submarines. From January 1, 1916, "U17" was placed at the disposal of the Submarine School.

"UB2" under Oblt. Hamm was transferred from Flanders to the Baltic Sea in mid-March 1916, but was unable to operate until the end of April.

"UB5" under Oblt. Smiths left Flanders for the Baltic in early October 1915; from October 30 to November 2, December 21 to 25, 1915 and January 1, 1916 the boat was in a holding position in front of Libau.

The new boat "UB20" under Oblt. Viebeg (later a prominent commander in Flanders) departed immediately after completing trials to the Baltic war theater. From April 16 to 22, Viebeg was in a holding position off Libau north of Kopparstenarne.

There are no noteworthy events to be found in this theater of war.

## Final Remarks

The third phase of submarine warfare wore an element of tragedy. This is the case not only because of the extraordinary restrictions that submarine warfare was subject to but especially because of the decisions that were made at that time. We must reiterate that if unrestricted submarine warfare had been declared in March 1916 and at the same time a plan adopted build submarines with intensive use of all German industry our success would be quite assured. Mistakes made later would have been made good by this timely commencement. It remains an open question whether conducting trade warfare according to Prize Rules in the months of May to October 1916 would have been sufficient to help a later, intensified submarine war succeed. (At the same time with a "large submarine construction program") There is no denying this possibility taking into account how close we came to decisive success in June 1917; the 1,600,000 tons of merchant shipping space not sunk in the months mentioned could actually have been decisive.

It was also tragic since the end of April 1916 the mandated, purely military use of North Sea submarines did not have a decisive influence on fleet operations under Admiral Scheer, especially in the Battle of Skagerrak. We will get to know the reasons for this when studying the fourth phase of the Submarine War.

# German Submarines and their Warfare 1914-1918

## Rear Admiral Albert Gayer

### Volume IV

### May 1916 to February 1917
### The Submarine Cruiser Warfare period and using North Sea Submarines against Enemy Warships

# Part IV - The period from May 1916 to February 1917



---

[139] These section heading is not boldfaced in the text as per the original.
[140] Ibid.

# Forward

Over eight years have passed since the publication of the third volume of my writings regarding the warfare of German submarines. It seemed appropriate to gain a greater distance from events before further publication.

The author is aware there is still a great deal to be corrected in the following narrative through further research of official publications in all countries involved in the events. He hopes, however, to have clarified important aspects of this very complicated warfare, which was heavily intertwined with politics, and to have contributed something by the following work.

The sinkings detailed do not make any claim to final and absolute accuracy either. The confirmed results are so marked.

Kapitän zur See Dr. H. C. Groos has reviewed this volume as well. At this time, I would like to express my special thanks for his very valuable assistance.

Hamburg, January 1930. The author.

# PART IV - The period from May 1916 to February 1917

## Introduction

In April 1916 it was an early summer over the waters of the German Bight when submarines began to gather at bases again after returning from their last trade war cruises or after completing shipyard repairs.

While using the boats for trade warfare required the departure of every vessel as soon as it was ready, "purely military use" meant - especially regarding airships – there was often long wait at the bases for suitable weather to commence operations. The wait had a particularly depressing effect on submarine commanders and crews. The spring weather, which was very suitable for operations on the west coast, invited one to fresh, happy deeds while military use tied the boats to the North Sea. Veteran submarine campaigners, who had been sailing against England for almost 2 years, no longer had any illusions regarding their chances of success against warships, if used against them without their own High Seas Fleet, no made how artfully they executed the search patterns and circuits the High Command made them search for the enemy. In fact, according to the Commander North Sea Submarines during the whole period there was only sunk the light cruisers "Nottingham" and "Falmouth" (August 19, 1916) and the cruiser "Hampshire" was destroyed by a mine, laid by "U75" (KptLt. Beitzen) on the west coast of the Orkney Islands (June 5, 1916).

In volume 3 it was mentioned September 1916 brought about a resurgence of commerce warfare in the North Sea, as well in that month three large boats were sent to the Northern Arctic Ocean to conduct trade warfare off the Murmansk coast. Another order shortly thereafter, dated October 8, 1916, ordered the general opening of the trade war according to the Prize Rules for the Flanders and North Sea submarines.

Accordingly, as far as the North Sea and Flanders war theaters are concerned, the fourth war period is dealt with in two sections:

1. The period from May 1 to early October 1916,
2. The period from the beginning of October 1916 to February 1, 1917.

In the Mediterranean a diverse use of the boats was not apparent. Trade warfare was waged there continuously. The activity of the Mediterranean submarines is therefore described in a uniform manner at the end of this volume.

## The Period from May 1 to early October 1916.

### A. North Sea.

The following boats joined the old, tried and tested ones:

April 29. "U51", KptLt. Rumpel,
May 1. "U46", KptLt. Hillebrand,
- 1. "U63", KptLt. Otto Schultze, -
- 8. "U47", KptLt. Metzger,
- 9. "U52", KptLt. Hans Walther,
- 19. "U75", Mineboat, KptLt. Beitzen,
- 30. "U53", KptLt. Rose,
- 31. "U64", KptLt. Morath,
June 7. "U48", KptLt. Buß,
- 28. "U76" Mineboat, KptLt. Bender,
- 28. "U77", KptLt. Erich Günzel,
July 1. "U54", KptLt. Frhr.v.Bothmer,
- 2. "U65", KptLt. v. Fischel,
- 9. "U78", Mineboat, KptLt. Dröscher,
- 26. "UB34",Oblt. Theodor Schnltzzn
- 29. "U55", KptLt. Werner,
- 30. "U79", Mineboat, KptLt. Jess,
August 6. "U49", KptLt. Hartmann,
- 17. "U56", KptLt. Lorenz,
- 17. "UB35", KptLt. Gebeschus.

- 26. "U80", KptLt. v. Glasenapp,
- 28. "U50", KptLt. Berger,

September 7. "U57", KptLt. Georg.

One can see from the strong boat increase how profitably these months could have been used in conducting trade warfare – even if only according to Prize Rules – because all of these 23 boats, except the 6 mine boats, were excellently suited to conduct trade warfare on the high seas outside of the well guarded zones.

Various positions were manned in the German Bight in early May. These cruises resulted in nothing new. They served as a certain security for our light surface forces, which often entered and left the area to escort airships.

## Jellicoe's Foray to Horns Reef
## May 2 to 5, 1916.

As we now know from Jellicoe's descriptions, the English fleet made an advance from May 5 to 5 in the direction of the Horns Reef to launch a seaplane air raid against airship hangars at Tondern. The English battlecruisers and battle fleet crossed on May 4 (according to Jellicoe) from 5 a.m. until 2 a.m. about 60 nm northwest of Horns Reef. The enemy was sighted there by "U24" (KptLt. Schneider), coincidentally, "U51", "U70", "UB21" and "UB22" were also at that time in positions which made directing these boats against the enemy seem promising. An attempt was made to place these boats on the enemy's track by wireless from the command ship, the cruiser "Arcona". In a short time it was possible to notify all of the boats mentioned of the enemy's location, unfortunately on "U24", the boat which came into contact with the enemy, a coupling between the diesel engine and the electric motor caught fire and it had to discontinue closing attempts due to insufficient speed. It was up to luck whether the other boats found the enemy; which was not in our favor. Neither did torpedo boats chasing after the enemy during the night; the English submarine "E31", which they reported as being successfully shelled, also escaped by luck. On May 8, "U32" returned from the northern North Sea, where it stalked enemy warships on the Mukkle Flugga (north point of the Shetlands) - Holmengroa (Norway) line. On April 30, he unsuccessfully attacked a Defense-class armored cruiser. On May 9, "U70" also returned without success. The boat had patrolled the Fair Island —Marsten line.

In the first half of May the efforts of fleet commander - Admiral Scheer - to strike a blow to the British took on ever more tangible form.

First, a plan was drawn up that would involve an attack on the English coast near Sunderland in order to pull the enemy out of their bases and strike them in front or south of the Firth of Forth. In preparation for this, a thorough reconnaissance by submarines of the North Sea between England and Norway was planned. After this reconnaissance, the boats should take waiting positions in front of the main English bases in order to attack exiting and returning enemy ships drawn out by our fleet. The boats were precisely distributed for this purpose in front of the Orkneys (Scapa Flow), the Firth of Moray, the Firth of Forth and the Humber; waiting positions at other important strategic points of the North Sea, such as at Terschelling Lightship, were provided by Flanders boats. It was a generous, carefully crafted plan. It was announced on May 13 by III and IV Submarine Half-Flotilla commanders to submarine commanders gathered in Emden on the accommodation ship "Rugia". The announcement was followed by a historically significant debate, which led to the following result:

1. A reconnaissance between England and Norway was of no particular value since the situation could fundamentally change shortly after the boats' departure from the reconnaissance positions as it was known the English frequently changed their so-called "cruiser areas" in the northern North Sea Blockade lines they continuously maintained.

2. The waiting positions in front of the enemy bases were also not expected to be very successful. Submarine commanders had

struggled in vain on this duty over the previous two years. Just like us, the enemy had learned to protect their combat forces particularly well when entering and leaving port, such that the prospect of submarine attack was kept to a minimum, unless night time was chosen for entering and leaving, as we almost always did, then the prospects of submarine attack were almost zero. It was also known the operation would wait until weather was suitable for aerial reconnaissance. It had been experienced many times before such waiting would usually extend beyond a week's duration. Submarine commanders could therefore expect a long stay in front of the enemy bases. Over such a long period and lying in such large numbers (7 boats alone in front of the Firth of Forth), they were bound to be discovered by patrols and hunted accordingly, which further exacerbated the danger to submarines and increased enemy security when entering and leaving harbour. It is interesting to read the Jellicoe confirms the correctness of this view. He writes that on August 18, 1916, the presence of an unusually large number of submarines in the North Sea - a phenomenon which had been shortly observed before the Jutland battle[141] - led him to suspect the German fleet was again intending to take action, and therefore special measures were taken.

3. That Admiral Scheer would definitely strike in the near future inspired all of the participants in the meeting and accordingly inspired them with the desire to be on the battlefield. The latter seems possible if the boats remain in their bases and sortie only shortly before the departure of the High Seas Fleet. With the large number of available boats (around 20), success seemed to be expected, especially if it was possible to approach the large enemy warships during battle, when there was no security against submarines. Unfortunately, such a possibility had never existed before.

Unfortunately the proposals made on the basis of this meeting with the Commanders of the III and IV Submarine Half-Flotilla to have

[141] Footnote in original. A phenomenon that had also been observed shortly before the Skagerrak battle (and even then gave rise to certain assumptions! The author.)

the boats depart shortly before the naval forces did, in the direction of the advance chosen by them, could not be taken into account since all dispositions had already been finalized.

Accordingly, the departure of the boats began as planned.

On May 13 "U74" departed on a mining operation to the Firth of Forth, it was tasked with throwing a field east and south-east of Bass-Rock. On May 16, "U24", "U32", "U63" and "U70" went to sea, and on May 17, "U43", "U44" and "U52", initially to their reconnaissance positions in the northern North Sea.

The new boat "U47" was made ready for with possible dispatch for the operation at the Ems base but despite all efforts, the boat left an oil trail; the oil bunker was sealed with the greatest difficulty. The boat departed on May 20, it was tasked with making a final in-depth investigation off Sunderland for our fleet regarding swept mine approach routes, lighthouse, sea trade traffic, etc. After reporting the reconnaissance results, the boat should wait in front of Kinnaird Head.

"UB21" and "UB22" went on May 21 to the Humber, "UB27" to the Firth of Forth to wait there between the coast and May Island to attack warships, "U72" to Moray-Firth to place a mine field between Stotfield Head and Kinnaird Head.

On May 22nd, "U46" and "U67" departed for a holding position at Terschelling Bank. The Commander the IV Submarine Half-Flotilla, KptLt. Prause, embarked on "U67" in the event an operation composing several submarines at sea had to be conducted. For nearly eight days these two boats skirmished with enemy submarines, which used the Terschelling Bank Lightship on their way to the German Bight. Destroyers and other fast vessels were also spotted by them. The personnel and material demands made on these two boats were particularly high.

On May 23 at 2 a.m., "U47" reported to the High Seas Fleet Commander the result of his reconnaissance off Sunderland; there were no concerns regarding navigating the fleet to this coastal point. Admiral Scheer, however, was unable to take advantage of this favorable circumstance, as the weather did not appear suitable until May 30th for airship reconnaissance. Admiral Scheer dared not dare sortie to the English coast without airship protection, considering in this case the benefits would lie with the enemy (fighting in his own waters) and the presence of the German reconnaissance boats only partly offset this inequality. From the submarines' perspective it was regrettable the raid on the English coast was postponed because the chances of intervening in battle would have been high. Admiral Scheer waited until May 30 to see whether the weather gods would still be favorable to him; for it was high time to do something with the fleet if the increase in submarines was to be put to advantage. It was therefore decided to depart north on that day towards the Skagerrak, instead of rushing towards the English coast, where, supported by the fleet, the light strike forces were to wage trade warfare.

Let us dwell briefly on the events preceding May 30.

On 28 May, "U72" reported he was returning. The mining operation was not carried out due to a failure of the diesel motor fuel pump. On May 29, "UB27" returned from its venture to the Firth of Forth. The boat had surveyed submerged Kirkcaldy Bay as far as Inchkeith and had seen no warships there, but on May 24th at North Carr Lightship, four older armored cruisers and destroyers were encountered; the boat was three times caught in nets and the starboard screw was obstructed. On May 30th at 8 a.m., "U46", entered the Ems, returning from its position at Terschelling, the only periscope in the boat had become unusable. KptLt. Hillebrand had therefore to give up his position there; he reported "at Terschelling Bank numerous floating mines, encountered many enemy submarines, and was attacked by enemy submarines once by torpedo, once by artillery". In view of the tense overall situation, "U46" was immediately sent to Emden to replace its periscope.

At midday on the 31st, the Ems Submarine Flotilla Commander had three boats ready for use, "U19", "U22", "U64". In Helgoland was

"U53". These boats had just come from the shipyard or had been released from the training cruises ("U53","U64").

The High Seas Fleet weighed anchor in the evening of May 30 to 31. On May 31, 7.40 a.m., "U32" reported 2 large warships, 2 cruisers and torpedo boats with a southeastern course from a position 155 nm east of the Firth of Forth. At 9:00 a.m., a message arrived from "U66", 8 large warships, small cruisers and torpedo boats in a position 60 nm east of Kinnaird Head with a north-easterly course. "U66" went in to attack this formation, but had not been able to shoot due to the strong escort.

From these reports, the first made at 6.10 a.m., the second at 7.53 a.m., it emerged that strong enemy forces were also at sea. The Commander of the III Submarine Flotilla immediately went from Borkum-Reede to the cruiser “Arcona” at Emden in order to, if necessary, use the stronger wireless there and personally direct the submarines at sea, according to instructions issued by the Commander of Submarines, then embarked on the cruiser “Hamburg”. At his request, the commander of "Arcona", KKpt. von Hippel, at 3 p.m. took the ship to Borkum-Reede where "Arcona" could communicate particularly well with the submarines.

At 4 p.m. came news from the small cruiser "Elbing" that 4 enemy small cruisers were in sight. It was clear then that the fleet had come into contact with the enemy as it advanced north. The three boats at Borkum-Reede, "U19", "U22" and "U64" were ordered to be ready to sail. "U46", which was in Emden, was instructed to clear harbor immediately as soon as the new periscope was installed and come to Borkum-Reede. At about 6 p.m., when our battlecruisers were also in action, the three submarine commanders were brought to "Arcona" and the war situation reviewed with them. If no special order was received, the intention was for the three boats to occupy the area between the Doggerbank barrier and the German Bight mined area, in order to intercept disabled vessels attempting to reach English southern ports after the battle (which was likely to have ended at nightfall). At 10 p.m., the signal from the Commadner of Submarines arrived at the Ems: "Ready submarines and

"U67"(which was at Terschelling) immediately head north. Location report at 6 p.m." Boats "U19", "U22" and "U64" left the Ems at 11:45 p.m. Transmission of the signal to "U67" proved impossible, as this boat, apparently heavily stressed due to a 14-day exposure in a dangerous area, was submerged.

From the messages received since 8 p.m. on the submarine wireless command ship, there was no longer any doubt on the F.T. that our line forces were engaged in battle. From midnight onwards, numerous signals were heard about the night battles underway north of Horns Reef. Although the 10 p.m. signal indicated the Commander of Submarines was still fully capable of radio-directing the boats, no signal was given to the boats lying in front of the enemy bases, given the great distance there was no prospect of getting the boats to the battlefield in time. The importance to Admiral Scheer if "U32" and "U66", given their speed, could have been able to follow the enemy forces and report need not be discussed further here. The whole deployment of our fleet for battle could have been based on reports from these boats.

At 6 a.m. "U46" left the Ems estuary to head north, as did the other boats; no one had any idea that this last-departing boat would have the best prospect of all submarines to attack a large enemy war ship. While the three boats now leaving were initially searching for the now sunken cruiser "Elbing" according to a Fleet Commander order received at 8.57 a.m., our radio directional stations reported at 10.30 a.m. a damaged English ship about 80 nm SW of Horns reef with course SW. "U46" being closest to this area was ordered to proceed on a course likely to intercept the enemy ship. Before this signal was received, "U46" sighted the enemy; it was the battleship "Marlborough", which had been hit by a torpedo in the battle the day before and was now enroute to a southern English shipyard, accompanied by only one destroyer. "U46" approached to 3000 m; the shot seemed certain, but the torpedo missed its target, since "Marlborough" noticed the launch and turned 8 degrees to starboard. Unfortunately the commander, in an effort to depart the battlefield, surfaced and continued northwards instead of continuing the attack or keeping in touch with the ship, which was only proceeding at 12 knots, because it was still possible to use "U67" and maybe place some Flanders boats on the course of the disabled vessel. "U67" tried to find the "Marlborough" on its own - without support from radio direction signals - but it did not succeed. Around about the same time "U51", KptLt. Rumpel, and later "U63", KptLt. Otto Schnitze, came across the damaged battleship “Warspite” between 40 to 80 nm east of the Firth of Forth. In the high swell both boats were unable to identify the ship as a battleship. The attacks were unsuccessful.

All other efforts to provide submarines an opportunity to attack were also unsuccessful. "U19", "U22" and "U64", under orders to advance to Peterhead and Scapa Flow, were so hampered by the onset of severe weather that they turned back; at 10 a.m. on June 2 they had stood at about the same latitude as the English formation, only further west. "U46" was sent to Flamborough Head by the III Submarine Flotilla Commander and sighted 4 small cruisers and 10 destroyers; the boat was not able to attack. No more successes were achieved by submarines against the incoming English ships either, since their entry into harbor took place on the night from the 2nd to the 3rd.

Thus, the Skagerrak battle proved unsuccessful for submarines; despite their huge efforts and lengthy periods the boats were exposed to off the enemy coast. The Flanders boats, positioned in such a manner to intercept strike forces leaving the Thames, were unsuccessful; they only sighted coastal patrols. After the battle, the opinion was widespread within the submarine force that if the boats were near the battlefield, which could have been achieved with appropriate deployment, the submarines would have been successful because enemy forces steamed through the same area repeatedly on reciprocal courses. It would have been necessary to send the boats out in such a manner that they were at Horns Reef along with light forces and be closer to the battlefield. However, at the focal points where the battle developed - this is to be admitted - it would not always have been easy for the submarines to orient themselves when submerged and distinguish between friend and foe with the heavy smokescreens, further the massive impacts of the heaviest projectiles, destroyers and torpedoboats heading for the attack at high speed, would probably have at these points forced some boats to bring themselves to go to greater depths for safety.

In between, however, there would have been excellent opportunities for them to attack in the vast area where submarines would have been distributed; especially during the deployment of the British Fleet, when many ships were unable to shoot; finally, there would be opportunities for such attacks during the return of the enemy north on the morning on June 1st. The maps Jellicoe included in his book very clearly show these possibilities. The mood in the submarine force only rose again when it became known on June 6 that the cruiser "Hampshire" with Lord Kitchener on board ran into mines west of the Orkneys laid by "U75", Commander KptLt. Beitzen.

After the Skagerrak battle, it was initially assumed the enemy would sortie into the German Bight to take advantage of the opportunity where many German ships were not ready to fight. Strong readiness was therefore maintained at the bases, and expected English warship courses, running from the Silver Pit or Horns Reef to the German Bight were extensively patrolled; in anticipation of increased movement of British forces between Scapa Flow and west coast ports, a submarine took up a position at the northern part of the Minch Passage. In the same period, submarines directed by our pilots, engaged in an intensified search of damaged and sunken ships in the area of the Skagerrak battle.

On June 14, the commercial submarine "Deutschland" went north from Helgoland to sea. Its commander, Kapitän König, was acquainted himself regarding war procedures used by submarines when entering and leaving the German Bight, as well as the type of English patrols in the North Sea and what the overall situation is as detailed in submarine commander reports and descriptions. Veteran submariners doubted whether it was easy to send such a large boat, which had only made training dives in the Baltic Sea, into the difficult conditions of the North Sea and recommended the boat should properly practiced at diving at sea before attempting a North Sea departure. This view was justified by the severe diving mishap "Deutschland" experienced north of Horns Reef while diving in a seaway. Only the presence of shallow depth allowed the boat and not dive further otherwise it would have been lost. The later loss of the second commercial submarine "Bremen" was attributed from the outset by submariners to such a diving mishap due to insufficient training; no value was attached to the fantastic rumors regarding “Bremen's” destruction by the enemy. To this day there is no evidence the enemy destroyed “Bremen”.

On June 14 “U28” left the Commander of Submarines’ control and went to the Submarine School in Kiel. In the second half of July, "U19" and "U22" were ordered transferred to the Baltic strike forces. Shifting of boats into school service or Baltic Sea indicated that the way in which the submarines were used in the North Sea did not necessarily seem to require the use of all the boats we had there, as submarine trade warfare would have done; therefore training and operational considerations have increased influence on the deployment of combat forces. When the boats were later brought back to the North Sea for the intensified submarine war, it turned out that they were not familiar with the conditions of warfare in the new theater.

In the period from July 3 to July 13, 14 boats led by the commander of the IV Submarine Half-Flotilla, KptLt. Prause, were in the North Sea to stalk enemy warships. Arranged in a circle with a 10 nm radius, the boats passed through the area between Norway and the Shetland and Orkney Islands. If enemy ran into this circle with strong forces a great success was expected. During this operation some boats engaged in trade warfare. "U48" under KptLt. Buß captured the English steamer "Pendennis" and sent it as a prize under the leadership of the Oblt. der Reserve Hashagen to the Ems, where the steamer arrived safely on July 8th. “U24”, “U46”, “U52” and “U69” sank three enemy patrol vessels by artillery fire in the Peterhead area. 31 prisoners were taken and 1 one gun was captured. "U53" (KptLt. Rose) sank the English steamer "Calypso" of 2,876 tons on July 10. On the return voyage, the boats met various English minelayers which were covered by light forces. Unfortunately, a torpedo attack by Rose on the enemy small cruiser "Galatea" did not lead to success.

On July 14, the recently launched "U51" boat, KptLt. Rumpel, was sunk by an enemy submarine 2 nm north of the Outer Jade Lightship. The torpedo hit amidships, only 14 men were able to save themselves in the boat's bow and stern compartments; five of them ascended from about a depth of approximately 20 meters. Four of them were recovered alive by patrol boats guarding the wreck site. "U51" was enroute from Helgoland to the Jade with raised radio masts, which naturally made it easier for the enemy submarine to attack because it facilitated fixing the boat's position. In the future, radio masts were only to be raised when the wireless arrester antenna could no longer be used to establish a wireless connection. (As was already the case with individual flotillas.)

In the period from July 28 to August 4 there was another patrol of about eight boats in double lines through the North Sea; this patrol also failed to succeed against warships. "U20" only sank the darkened English steamer "Aaro" (2,603 tons) off Stavanger on August 1, which he took to be an auxiliary cruiser by its behavior. The rescued crew was brought to the Ems by "U20" and "U69".

The main boats tasks at this time heavily used both personnel and material: Occupying outpost positions, patrolling certain border areas of the German Bight, acting as reception points for airships.

On August 8 "U47" (KptLt. Metzger), under Naval Staff orders, began a transfer to the Baltic Sea Forces (see Baltic Sea, p. 290).

When the fleet advanced on the English east coast on August 19, the experiences of the Skagerrak battle regarding the use of the submarine were fortunately taken into account. The boats departed shortly before the fleet, on the night of August 17-18, to be positioned in two standing lines only on the morning of the actual day of the operation, one in front of Flamborough Head (4 boats), the second on the height from Blyth (6 boats). Currently, it is known that the commander of the British fleet had information regarding the presence of numerous German submarines in the North Sea. On this basis, he concluded the German fleet was planning to undertake some sort of operation, and accordingly departed Scapa Flow the evening of August 18. At dawn on August 19 the British fleet was at the latitude of the Long-Forties. German boats completed the trek to their positions in such a manner so as to sweep the greatest possible sea area for the following surface fleet.

At 7 a.m. on August 19, light cruisers screening for the English battlecruisers hit the submarine line in front of Blyth, whereby "U52" (KptLt. Walther, Hans) succeeded in sinking the modern English small cruiser "Nottingham" with three torpedo hits. Soon Kptlt. Rose, "U53", came into contact with the main enemy forces and reported its location. Unfortunately, these reports only reached Admiral Scheer when he was already prompted by an airship report ("L13", 1:00 p.m.) to abandon his actual plan to advance against the English coast and to turn against an enemy force, which "L13" sighted further south on the latitude of the Humber. The volume of our work cannot go into more detail why Admiral Scheer based further operation on "L13's" message; it should only be said here that "L13" gave incorrect information as to the number and type of enemy forces sighted (it was "Harwich force" under Commodore Tyrwitt), and due to ignorance of his own exact location also provided an imprecise location for the enemy, so that Scheer could not find them, but even if he had succeeded, "Harwich force", consisting of light forces, would have been easily evaded our attack.

According to Jellicoe, "Harwich force" gained contact with our fleet heading homewards at 7 p.m., but refrained from launching night attacks by destroyers since the conditions seemed unfavorable for this and the gloomy experiences of English destroyers at the Skagerrak battle probably still had an effect.

In the first hours of the afternoon, the boats of the southern submarine line also saw smoke clouds from enemy forces, "U49" and "U66" went on the attack. The situation was unclear for "U49" because Admiral Scheer's decision (1.15 pm) to turn south was not known. When "U49" had a small cruiser with four funnels in front of him at around 5 p.m., it was initially thought to be a German one. Only when an aft tripod mast was seen at about 5000 meters was the ship recognized as belonging to the enemy, v. Bothmer ("U66") succeeded in attacking the English small cruiser "Falmouth" at 6 p.m., which he hit with two torpedoes; nevertheless, the ship did not initially sink; "U66" and "U49" sought to complete the sinking before nightfall, which was not possible due to good protection by three destroyers. Only the next day (12.55 p.m.) when "Falmouth" was under tow to Flamborough Head, protected by seven destroyers, did KptLt. Schultze (Otto) with "U63", manage to sink the ship by two more torpedo hits. "U63" was rammed by one of the enemy destroyers during the attack, but fortunately with slight damage.

In the same evening (8.50 p.m.), "U65" under KptLt. von Fischel and "U44", Wagenführer, who belonged to the northern sentry line, attacked the English fleet returning northward. Two torpedoes from "U44" passed close to the stern of the battle cruiser "Inflexible".

This operation was the most successful undertaking the surface fleet and submarines have ever carried out together, despite a number of errors due to limited experience in joint activities. Jellicoe also recognized this venture, which developed from a "submarine trap", as operationally feasible and expresses wonder why this type of combined fleet and submarine use was not attempted as often as possible in the first years of the war. In this respect, Jellicoe's astonishment coincides with the numerous wishes and suggestions made in vain the first two years of the war by the Commander of North Sea Submarines, FKpt. Bauer. All submarines returned safely to base by August 22.

Subsequently, submarines were often used to provide cover for our airships, which were very active in attacks against England at the time. This use ultimately had the effect of creating special zones off the English coast, from the north exit of the Minch to Flamborough

Head (where the operations area of the Flanders boats began), occupied by submarines; it was believed the boats would be more successful if airship attacks increased the activity of warships off the English coast. The last such operation was launched in late September. There was no success against warships; on the other hand, the boats began to renew trade warfare more or less on their own. ("U49", for example, sank three sailing vessels and sent a Norwegian steamer as a prize to the Ems, "U57" sank a larger steamer and 21 fishing trawlers in front of Flamborongh Head, "U55" (KptLt. W. Werner), assuming a waiting position against warships in front of Pentland Firth on September 20m, sunk the English fish steamer "Orsino" (172 tons) and brought the Swedish steamers "Atle" and "Talavera" as prizes. Trade warfare was already in the air again; on September 20, 1916, "U43", "U46" and "U49" departed to the northern Arctic Ocean to conduct trade warfare according to Prize Rules. As early as September 17, "U53" had left Helgoland for a voyage to America in order to take action against enemy armed forces, which might guard the entrances to Long Island Sound in anticipation of the arrival of the commercial submarine "Bremen". This boat, too, had been permitted to wage trade warfare according to Prize Rules after completing its specific task.

The order for the general commencement of the trade war according to Prize Rules was received at the High Seas Fleet on October 8; using boats for purely military purposes was interrupted for two long years. Only in October 1918 did this operative idea briefly revive, when submarine trade warfare was abandoned to initiate ceasefire negotiations based on Wilson's 14 Points[142], only to be buried with the general moral defeat of Germany soon afterwards.

If the period of "purely military use" of North Sea submarines (May-October 1916) is not marked by any major successes, it should be

[142] The second point addressed submarine warfare. To quote, "II. Absolute freedom of navigation upon the seas, outside territorial waters, alike in peace and in war, except as the seas may be closed in whole or in part by international action for the enforcement of international covenants."

attributed to the lack of established joint operations between the surface fleet and submarines; a corresponding use of the boats would to a certain extent have served as a replacement for airship reconnaissance, the latter was usually waited on for weeks and then caused much confusion through erroneous reports and occasionally created mistakes in the management of operations. In some cases – the Skagerrak battle - the method of using the boats was tactically unfavorable and they had no success. It should be emphasized the use of submarines with the fleet was well and properly practiced in peace; had this path been taken further in the war it would have shown the submarine is also capable of playing an important role in naval combat, both in submerged daytime attacks and in surface night attacks. The slow speed of most boats was certainly a hindrance to such use at the beginning of the war, by the Skagerrak battle we already had a large number of boats that could at least temporarily run 15 to 16 knots, and by sending this boats out with the light striking forces almost always guarantee they could be available for the surface forces if they entered battle; especially owing to the enemy's superiority, the general course of our forces in a battle would mostly be to our bases, which is why any lagging boat is automatically in an attack position. The Commander of North Sea Submarines repeatedly advocated for submarines in joint operations the need to allocate fast destroyers, equipped with underwater signaling equipment and special radio installations for submarine traffic, but unfortunately his demands were not met.

During the period described, English submarines began to plant their mines closer to our estuaries, following the pattern of our Flanders boats. Even the crossing from the Ems to the Jade, which was mostly close to the 10 meter line, was considerably disturbed for a while at the end of July by a mine barrage found in the area of the Juister Reef howling buoy. Since the howling buoy naturally made it easier for an enemy submarine to control its navigation, it was soon replaced by a small spar barrel.

In technical terms, tests with fog bombs and cable cutting drags for submarines had a major role in the period described.

## May 1916 to February 1917

The activity of the large mine submarines "U71-80", which operated under the direction Commander of North Sea Submarines was considerable in the period described; and can be seen from the following compilation.

| Boat | Cdr. Name | Sortie Date | Minefield location | Remarks |
|---|---|---|---|---|
| U71 | KptLt. Hugo Schmidt | June 6 to July 10, 1916 | West of Cape Barra Head and West of Skerrymore Island | West of Scotland 48 mines |
| U72 | KptLt. Krafft | May 23 to 29, 1916 | It was not possible to lay mines at the Firth of Moray due to fuel leakage and damage to the pumps. | |
| | | June 6 to July 4, 1916 | June 27th mine field planted on a course between Pentland Firth and Cape Wrath | |
| | | August 20 to Septem ber 17, 1916 | Mines planted on September 3 at Lisbon, September 7 at Oran, September 11 at Bizerte | Voyage from Helgoland to Pola. |
| U74 | KptLt. Erwin Weisbach | | Departed May 13 to lay a mine field at Bass Rock at the entrance to the Firth of Forth. Surprised by the admiralty drifters "Searanger", "Oku" and "Rodino" "when traveling surfaced and destroyed by artillery (100 nm from the coast of Aberdeen) | |
| U75 | KptLt. Beitzen | May 24 to June 3, 1916 | May 29, 1916 mine field planted west of the Orkneys Islands in the morning | On June 5 the English amored cruiser "Hampshire " ran onto these mines and sank with Lord Kitchener on board. |

| Boat | Cdr. Name | Sortie Date | Minefield location | Remarks |
|---|---|---|---|---|
| U75 | KptLt. Beitzen | July 23 to August 2, 1916 | August 4 At the entrance to the White Sea, north of Cape Orlov. | |
| U76 | KptLt. Bender | July 26 to August 2, 1916 | July 30- off the English coast, north of Longston | |
| | | Septembe r 17 to October 18, 1916 | October 2 White Sea, to the NW from Cape Gorodetsky, October 3 at Cape Holy Nose. | In the White Sea |
| U77 | KptLt. Erich Günzel | July 5, 1916 left Helgoland to lay mines on the route from Kinnaird-Head to Klock Head off the Firth of Forth. | | Boat did not return. Cause of loss unknown |
| U78 | KptLt. Dröscher | July 13 to 28, 1916 | July 21 in the area between Barra Head and Skerryvore. South entrance of Hebrides Sea. | A steamer was taken as a prize off Scottish west coast and sent to Germany. |
| U78 | KptLt. Dröscher | August 20 to Sept. 12, 1916 | September 1, on course between Hellwick Fisch - St.Gowan Fisch | At the northern exit of the Bristol Channel. |
| U79 | KptLt. Jess | August 6 to 25, 1916 | August 17 - Off Courtmacsherry, Clonakilty and Glandore Bay | Off Irish southcoast, west of Cape Old Head of Kinsale and west of Galley Head. |
| U79 | KptLt. Jess | Septem ber 26 to October 14, 1916 | October 4th between Sanda Isl. And Ailsa craig lighthouse and Pladda lighthouse | North of Irish North Channel. At the entrance to the Firth of Clyde. |
| U80 | KptLt. v. Glasenapp | Septem ber 10 to 18, 1916 | September 12 on a course between Farn Island and Coquet Island | English Eastcoast |

In addition to the cruiser "Hampshire", the following English steamers fell victim to these mines:

The drifter "Laurel Crown" on June 2, 1916, which sank through mines west of the Orkneys according to the English white paper "Navy losses". Only the mines laid by "U75" west of the Orkneys on May 29 are the possible ones. It is a testament to the extent to which the English naval leadership was shaken by the result of the Skagerrak battle that Admiral Jellicoe had "Hampshire" take the route north west of the Orkneys, although, as Jellicoe expressly observes, a search for mines due to heavy weather was not possible.[143] The loss of the "Hampshire" and the Field Marshal is indirectly a consequence of the Skagerrak battle. One must assume the loss of "Laurel Crown" on "U75's" mines was not known to Admiral Jellicoe when he wrote his memoirs.

Further successes for Beitzen were the sinking of the trawler "John High" of 228 tons on August 7, 1916 before Sozonova and the loss of the collier “Etton”, 2831 tons, on September 20, 1916, both in the White Sea.

A clear success for Dröscher ("U78") is the sinking of the steamer "Kelvinia" (5,039 tons) between Helwik and St. Goven Lightship. Dröscher laid his mines there the day before on September 2nd.

The sinking of the "Counsellor" (4,958 tons) 5 nm W1/2S from Galley-Head is due to the mines laid there by Jeß ("U79") in August.[144]

143 A rare lapse in judgment by Gayer. “Hampshire” used the NW route due to the prevailing extremely heavy weather.

144 Footnote in original. The sinking of the auxiliary cruiser "Laurentic" (14,892 tons) on January 2, 1917 is likely to be due to one of "U79's" mines on October 4 at the entrance to the Firth of Clyde or to one of "U80's" mines on October 27 between Isle of Man and Holy Head.

## Flanders

In Flanders the situation was one of stagnation for the entire period from May to October as the Commander of the Flanders Submarine Flotilla sought to comply with the operations cessation order of April 25, 1916, extending it to the new torpedo and CII boats. First, the net barrage laid out by the enemy in April off the Flanders coast brought work to the submarines. It was determined the guard mantianed by the enemy for this barrage was based in Dunkirk. The approach to Dunkirk was therefore systematically mined by C-boats; it was also decided to keep two torpedo submarines in front of the barrage as outpost boats. These boats extinguished the lights of the marker buoys put out by the enemy at night for the protection of the boom network. When it became clear the enemy consistently re-ignited these buoys, on May 12 Oblt. Wenninger with "UB17" was tasked with destroying all these buoys which was completed by May 14. When it became clear the net barrage was not serving a purpose for preparing an attack on the Flemish coast and intended only for anti-submarine warfare, it was decided to no longer hold back the big type BII boats and instead deploy them against enemy forces off the Humber river mouth and in the Hoofden. Coast guard duties devolved to the small (BI) boats. The outpost positions of these boats were generally taken east and west of Thornton Bank. There was the further task of guiding boats coming from German harbors to Zeebrugge. On May 19, "UB23" arrived in Flanders under the command of Voigt. It was a replacement boat for the "UB26"[145] lost on April 5 at Le Havre (Oblt. Smiths). At this time Oblts. Steinbrink ("UB18"), Becker ("UB19"), Pustkuchen ("UB29") had various skirmishes with English fishing trawlers, some of which were armed and unexpectedly opened fire against submarines or as part of the enemy patrol service; all of the trawlers were sunk.

On May 30th and 31st, 8 Flanders torpedo submarines and 3 mine boats went to sea to support the operation of the High Seas Fleet which resulted in the Skagerrak battle. The idea was to intercept

---

[145] Footnote in original. Volume 3, page 146.

enemy strike forces on their way to and from the Thames with torpedoes or mines, in accordance with the plan also utilzing North Sea submarines. Attempts were made to achieve this goal by movable submarine lines. This method was unsuccessful for the Flanders boats as described previously in the section dealing with the Skagerrak Battle course of events. It would have been correct to use the four modern Flanders boats "UB18", "UB19", "UB23" and "UB29" in joint operations with the High Seas Fleet, in the same way as was already required for North Sea submarines.

The English measures - laying out net and mine barriers against the Flanders boats - made it necessary to move the Dutch lightships "Nordhinder" and "Schouwenbank" further north shortly after mid-June. This made it even more difficult for submarines to approach the Flanders coast.

"UB39" arrived in Flanders on June 21 under the command of Oblt. Werner Fürbringer.

At the beginning of June the II Torpedoboat Flotilla was dispatched to Flanders under the command of the FKpt. Schnur, to take action against British net tenders and anti-submarine vessels and so provide greater freedom of action for Flanders Flotilla submarines. In operations this flotilla made from Flanders the submarines often served as guide boats.

On July 5, "UB37" under the command of Oblt. Hans Valentiner arrived in Flanders from Germany as further reinforcement. During the crossing the boat radioed its position as further west than it actually stood; it was mistaken for a hostile submarine by Ems guard steamers, bombarded with artillery and depth charged. The "enemy submarine" was reported by the Ems patrols as destroyed, and the facts were only clarified shortly before awards for its destruction were to be awarded. There was some consolation as the Ems patrol was praised for its persecution of "UB37"; it was perceived as very appropriate and persistent.

On June 29, "UB19" and "UB23", and on July 6, "UB18" and "UB39" were dispatched to be in position off the Tyne estuary with the task of attacking enemy warships, especially after the Skagerrak battle repaired ships could be intercepted when leaving there. Enemy forces were sighted several times, but there were no opportunities to fire. "UB23" sank 5 English fishing trawlers and 3 fishing cutters (478 tons), "UB18" 2 sailing vessels and 6 fishing trawlers (approx. 600 tons). "UB39" bombarded the Seaham Steelworks with 89 rounds on the evening of June 11 and also sank 8 fishing trawlers and 2 fishing cutters (1,074 tons). A similar operation for the Tyne dispatched Oblt. Hans Valentiner with "UB37" from July 15th to 24th. On July 18 the boat got caught in a net, with the tower and periscope damaged; 5 Norwegian and Swedish sailing vehicles (1,767 tons) were sunk, a steamer was sent to Emden as a prize. As can be seen here too commercial warfare activity inevitably developed from operations intended only for warships.

In mid-July, the Chief of the General Staff demanded submarine mines disrupt enemy troop transports to France.[146] The compilation given below describes how the Commander of the Flanders Submarine Flotilla attempted to meet this request by repeatedly laying mines in front of Calais; the Commander of Flanders Submarines, of course, made no mistake that an effective disruption of Channel transports was only possible through (prohibited) torpedo attacks without warning; He expressed this conviction in reports to the Naval Staff.

On July 21st, the submarine exit route leading north from the Dutch sands of Zeebrugge was blocked by mines. The mine clearing flotilla was fired on by British cruisers, monitors and destroyers to disrupt sweeping. It was not until July 22nd that a swept passage was created.

From July 22nd to August 1st "UB23" - Oblt. Voigt - was in front of the Tyne. 1 lighter, 2 neutral sailing vessels, 8 motor sailers and 1

---

[146] The Battle of the Somme was undeway at this time.

fishing trawler (together 267 tons) were sunk. At that time countermeasures against submarines with depth charges began to increase. "UB23" was the recipient of two such bombs.

Since mine warfare, as mentioned, was insufficient to disrupt the transports, "UB18" under Steinbrink was sent on July 30th to the Channel. He was only to shoot down positvely identified transport steamers, recognizable by troops or military equipment visible on deck; in all doubtful cases, Steinbrink was to refrain from attacking. This "test task", which was initially carried out before the confluence of the Seine and then off Le Havre, clearly showed it was not possible to seriously impede transport traffic with such procedures. Steinbrink made a total of 41 attack approaches, but the attacks had to be stopped before firing because the conditions specified in the instructions for identification of transports were not satisfied, only 6 smaller steamers, a small tanker and 10 sailing vessels (together 5,500 tons) were sunk according to procedures in accordance with Prize Rules.

On a cruise from July 29 to August 7, 1916, Oblt. Werner Fürbringer with "UB39" sank 3 English steamers and 10 English fishing trawlers of 2,042 tons. Oblt. Becker with "UB19" sank the English steamship "S. Bernardo "(3,803 tons) with explosive cartridges 17 nm southeast of Longstone. When on August 7, 1916 the Chief of the Naval Staff v. Holtzendorff came to Bruges for briefings with the Flanders Command, KKpt. Bartenbach tried again to obtain permission for vessels on courses N and S, bounded by the Cape de la Hague-Portland and Calais-Downs lines should be sunk without warning, and the operations area before the Tyne to be finally allocated to Flanders because of the good results achieved there. The Naval Staff replied on August 15 in the negative to this request, explaining the situation on the Somme did not pose such an emergency to risk a rupture with America which torpedoing assumed transports in the Channel without warning must create. "UB10" under Oblt. Buch sighted several enemy destroyers on a cruise along the England-Holland steamer routes on August 13; he managed to sink the destroyer "Lassoo" (1,010 tons).

To support an operation by the High Seas Fleet on August 19, a submarine line was laid out from Flanders approximately midway between Terschelling and the Humber with "UB18", "UB23", "UB29" and "UB39", another with "UB6 "," UB12 ", UB16", "UB19" and “UB37" at Terschelling. These boats had no encounters with enemy strike forces. Unquestionably it would have been advantageous to push them further north to the latitude laid down by the High Seas Fleet for submarines. The Flanders boats were less experienced in the use of radio telegraphy, given their different role in the Channel, was also a major drawback in joint action with the surface fleet.

Since all attempts by the Commander of the Flanders Flotilla to give his boats broader work against transports in the narrower part of the English Channel failed, he decided to send the more powerful new submarines to the western part of the Channel and conduct trade warfare according to Prize Rules in order to, as he put it, not have the boats lay idle. In fact, he was doing what the Naval Staff had authorized since April, but did not categorically demand. Subsequent cruises demonstrated what successes could be expected from the Flanders boats. It must be taken into account the results listed below are for the initial phase of operations, and these successes could not have been sustained at the same level if work had begun from May onwards.

On September 1, 1916, "UB18" - Steinbrink -, "UB23" - Voigt -, "UB29" - Pustkuchen - and "UB39" - Werner Fürbringer - left for the western part of the Channel. There they acheived the following sinkings:

Steinbrink: 12 large, 3 small steamers, 5 sailing vessels, altogether: 36,879 tons, eight of them enemy vessels of 5,124 tons, the rest neutral (Sweden, Norwegians, Greeks, Spanish, Danes and Dutch).

Voigt: 2 steamers, 4 sailing vessels, 4 fishing trawlers, totaling 6,077 tons, including 1 neutral steamer of 1,880 tons.

Pustkuchen: 3 steamers, 2 sailing vessels, 5 fishing cutters totaling 8,224 tons, including the English steamer "Torridge" (5,036 tons) 40 nm SSW from Startpoint and 2 neutral steamers, including 1 Dane of 1,191 tons and 1 Norwegian of 1,463 tons.

Werner Fürbringer: 7 steamers, 4 sailing vessels totaling 19,661 tons, including 3 neutral steamers displacing 5,097 tons.

Overall, this initial operation carried out essentially in the western end of the English Channel, brought a result of 68,313 tons; the Commander of the Flanders Submarine Flotilla therefore decided to continue trade warfare according to Prize Rules, although severe damage to enemy shipping did not appear to be very promising because of their strong protection and defensive armament, and attacking military transports in this manner was naturally impossible. The low surface speed of the BII boats - just under 9 knots- and their armament being only a 5 cm cannon naturally made this type of activity more difficult for them. Up-arming the boats to 8.8 cm-S.K.[147] was immediately adopted. While the war-experienced BII boats operated far away from the Flanders coast, the small BI boats and the new BII boats coming from home worked on this side of the Dover-Calais line in the Hoofden, where also some sinking using Prize Rules were achieved and various ships were brought in as prizes.

On September 11th, "UB38" arrived in Flanders from Germany under KptLt. Waßner. After Waßner had carried out a patrol on the Dutch coast between the Meuse and Haaks lightships in mid-September, he also cruised from September 27th to October 8th conducting trade warfare in the western Channel, where he sank 6 steamers and 4 sailing vessels of 6,198 tons in total occasionally using only artillery fire.

This period of voluntary participation in trade warfae on the basis of Prize Rules in Flanders is completed by the actions of the following commanders:

---

[147] SK = Schnelladekanone (quick loading cannon)

From September 21 to 28, Hans Valentiner sank 2 English and 1 French steamers and 1 French fishing trawler (3,260 tons) with "UB37".

Oblt. v. Heydebreck ("UB6") sank 1 neutral steamer on September 10 and 5 Belgian barges (total 1,086 tons) on September 23.

Oblt. v. Werner ("UC16"), who arrived from home with this first of the CII boats on September 11, sank 11 English fishing trawlers totaling 1,605 tons after a mining operation from September 21 to 24.

Oblt. Graf v. Schmettow arrived in Flanders with "UC26" on September 12th and went on a mining and trade warfare operation on September 28th (see page 240) to the western exit of the English Channel. Two enemy sailing vessels, one English and two Norwegian steamers were sunk totaling 5,518 tons.

Oblt. Becker with "UB19" sank 3 English fishing trawlers and 1 motorized fishing vessel (613 tons) on October 4 and 5.

Thus, October 8 order to open a general trade war on the basis of Prize Rules (submarine cruiser war) found the Flanders Flotilla already fully practicing this activity; there is also no doubt this operation, voluntarily undertaken by KKpt. Bartenbach due to the lack of better use of the boats, contributed significantly to the issuance of this general order.

The following description shows the activity of the Flanders mine boats in the period from May 1 to October 8, 1916:

"UC1" under the command of Oblt. Ramien 12 sorties, under Oblt. Küstner 4 sorties in minelaying operations (16 X 12 - 192 mines). One of these cruises was directed against Dunkirk, one against Calais, all other mines were carried to the English east coast. The initial minelaying took place beyond the Dover-Calais Narrows.

"UC3" laid mines on the approaches to Dunkirk again in May (11th to 12th) under the leadership of the Oblt. Waßner, then the boat came under the command of the Oblt. Kreysern. The boat laid mines at Dunkirk on two more ocassions (on May 14th and 21st). On May 25th it departed to lay mines in front of the English east coast (Aldbro' Napes Buoy), but did not return from there. It was sighted at the Nordhinder lightship on May 26 by two other Flanders submarines enroute home.

"UC6" planted its mine load 15 times under the leadership of the Oblt. Ehrentraut and 8 times under Oblt. Günther. All made trips to the Thames estuary or the English east coast north of the Thames except for two sorties by Oblt. Ehrentraut who were directed against Calais (July 2nd and 18th).

"UC7", under the command of Oblt. Haag, planted six more minefields in front of the Thames estuary. It did not return from a sortie that started on July 3 to mine the Elbow Buoy south of the Thames estuary. The commander's body was found and buried on July 24 by "UB16". It is highly probable "UC7" was torpedoed by an enemy submarine.

"UC10" sortied 9 times under Oblt. Nitzsche to mine the English coast, 1 time to Calais. On July 24, it had battle with an English aircraft where the latter was forced down. The airplane and 2 officers were brought to Zeebrugge. With the same boat Oblts. e.g. Saltzwedel and Albrecht each made 2 mining sorties to the English coast. Albrecht did not return from a sortie made on August 17, 1916 to the Humber, although after completing its task and sinking an English fishing trawler, it was sighted on August 21 by "UB10" at the Nordhinder lightship. According to English reports, "UC10" was attacked and destroyed with depth charges from the English motorboat "Salmon".[148]

"UC11" carried out 6 mining operations to the Thames estuary between August 8, the date when the boat arrived in Flanders from the Submarine School, until 8 October, the first trip under the

[148] Modern research has placed “UC10” as being sunk on another date by a British submarine.

command of the Oblt. Saltzwedel, the other 5 under Oblt. Max Schmitz.

In addition to these "old warriors" mentioned, the significantly improved mine boats "UC16" (Werner) and "UC26" (Graf v. Schmettow) appeared in Flanders during this period. Werner planted his mines - now 18 instead of 12 for these boats compared to the first C boats - in front of Hewett Channel, the southern entrance to Great Yarmouth; Count v. Schmettow, during the night from September 18 to 19, laid mines north of the Outer Dowsing Lightship on the approaches to Hull and Grimsby. In a subsequent trip, he passed through Dover-Calais and laid mines off Boulogne, Le Havre and Cherbourg on the 29th and 30th. Comparing both boat types we see the features allowed far more extended operations with these modern type CII boats compared to CI boats; displacement increased from 168 tons, to around 420 tons. Two 250-300 horsepower oil engines compared to the CI boats one 90 horsepower oil engine, which gave them almost twice the speed of the CI boats (12 knots to 6 knots); the range of action was about 10,000 nm versus about 800 for the CI boats. In addition to 18 mines, the boats carried 3 torpedo tubes, seven 50 cm torpedoes and an 8.8 cm quick firing cannon. The boats were a particularly successful design and a great achievement of our submarine construction.(see Fig 19.)

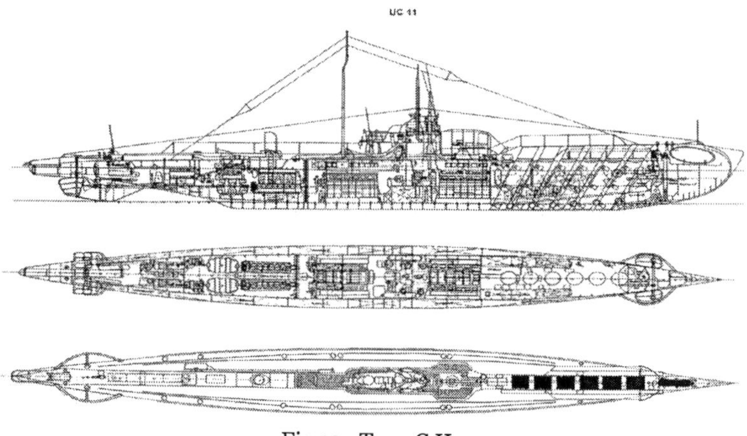


Fig 19. Type C II

Flanders boats planted 618 mines during the period described.

Mine successes May 1916 were: 10 vessels of 21,905 tons sunk, including 8 English, 1 Dutch, 1 French steamer; in June: 16 vessels of 18,794 tons, including 10 English, 2 Dutch, 1 American, 1 French, 1 Spanish, 1 Norwegian vessels; in July: 7 vessels of 10,958 tons, 5 British, 2 Dutch; in August: 1 English steamer and 4 English fishing vessels of 1,380 tons; in September and until October 8: 6 English steamers, 3 English fishing trawlers, 1 Norwegian steamer totaling 13,472 tons.

Total mining success of this period was therefore 66,509 tons. Information provided is based on official data drawn from White Books.

## Mediterranean Sea

In the Mediterranean it is not necessary to separate into distinctive parts the period from May 1, 1916 to February 1, 1917; because, as mentioned, there was no cessation of activity. As KKpt. Kophamel recognized with certainty that trade warfare according to Prize Rules was possible and conducted there continuously. (See Volume 3, page 157). Therefore, the activity of boats in the Mediterranean will be described in a single sketch at the end of this volume.

## The period from October 8, 1916 to the commencement of unrestricted submarine warfare (February 1, 1917)

## North Sea

The deployment of "U53" to America, and the order issued on September 15 for three submarines to be sent to the northern Arctic Sea to conduct trade warfare revived the stagnation of the submarine trade war in the North Sea from the dismal "purely military use" of the boats.

"U53" departed on this trip on September 17; primarily to attack enemy forces believed to guard the entrances to Long Island Sound against the commercial submarine "Bremen". - "Bremen" sailed at the beginning of September, but never reached its destination apparently as a result of a diving accident. It has been established that no enemy strike forces or mines did have a role in its loss. Enemy forces were not found when "U53" arrived on October 5 off the American coast. On October 7, "U53" entered the port of Newport. The same day after the a visit from the American station chief, Admiral Knight, and after a boisterous welcome by the sensational American population, the ship departed again and sank on October 8, in the area around the Nantucket lightship, the English steamers "Strathdene" – 4,321 tons, "Westpoint" – 3,847 tons, "Stephano"–3,449 tons, the Dutch steamer "Blommersdjik" – 4,835 tons - and the Norwegian steamer "Chr. Knutsen "- 3,936 tons, together 20,388 tons, sunk. The sinking took place partly in the presence of 16 American destroyers, which rescued the crews of the sunken ships. On the same day, "U53" had to start the return voyage due to a scarcity of supplies - no stopover had taken place in Newport; with skillful use of the Gulf Stream, which was reached on October 10th, the return journey went smoothly, and on October 25th Rose moored safely in Helgoland. This operation is one of the finest achievements of the German submarine war.

The III Submarine Flotilla was assigned to carry out the operation in the Northern Arctic Ocean. "U43", KptLt. Jürst, "U46", Oblt. Hillebrand, "U48", KptLt. Buß, were selected. Only the experience of "U75" existed about operating submarines in the area, which under KptLt. Beitzen had placed a minefield there on August 4th in the entrances into the White Sea in front of and north of Cape Orlov.[149] For the torpedo submarines a longer stay was needed in the operational area and it was doubtful how this would work out in the already advanced season; in this regard all concerns were dispelled by the very favorable experiences. They actually found a kind of "Riviera" in the far north. As a general plan, the boats were given the task of first loittering at North Cape area for steamers leaving the Lofoten archipelago and then working on a line running

[149] Footnote in original. The experience of "U76" (Bender), who returned from another mine operation in the White Sea the day before the three 40s boats left (October 18), was not yet available to the III Submarine Flotilla when the operational command was being drafted.

northwards from Vardö. The latter position also had the advantage that rescued ship crews could be brought with the support of submarines to this Norwegian port, which was furthest to the east and subsequently proved to be useful for the demand that trade warfare should be conducted in the most considerate form "according prize rules". In the event that joint operation of the boats was an option, KptLt. Jürst, as the oldest commander, was entrusted with the responsibility, which he handled with great skill. On October 19, the boats - excited about the resumption of promising activity - departed the Ems to the sounds of the ship's band of the coastal armored ship" Beowulf". The operation was particularly successful due to the lack of countermeasures in this area, the good weather and the careful handling of the boats. KptLt. Buß after sinking 3 steamers (one of them "Dania Norge"[150] had already been hit by "U43) took the Russian steamer "Suchan" (the German Hapag steamer "Spezia" taken at the beginning of the war in Vladivostok) with ammunition, guns and automobiles worth 40 million gold marks away off North Cape and brought this prize to Wilhelmshaven from October 8th to 16th. The English blockade was avoided with great skill. The commander of the prize, Oblt. Hashagen, watch officer on "U48", showed great energy and diligence, especially after experiencing stormy weather, scarce coal and the fore-ship partially flooded due to a shell hole, the prize arrived together with "U48" on October 16 in Wilhelmshaven. Jürst ("U43") and Hillebrand ("U46") extended the operation longer. The former sank a total of 10 steamers (25,344 tons) and fired at the telegraph station at Cape Tzuip Navalok on the fishing peninsula (Ribachin) in the night from October 7th to 8th, Hillebrand brought down 8 steamers. Both boats successfully fought various battles with Russian torpedo boats, which withdrew from the fire of the two submarines. With the arrival home of "U43" on October 21st and "U46" on 24th the operation came to an end, which brought in 55,013 tons of ship's displacement including the prizes.(see Fig 20. for Sept. Oct. Nov 1916 Artic Ocean success.)

---

[150] Actual name is "Dania". Norge is to signify Norwegian as neutrals had their nationality painted on their hulls in oversized letters to announce their neutrality.

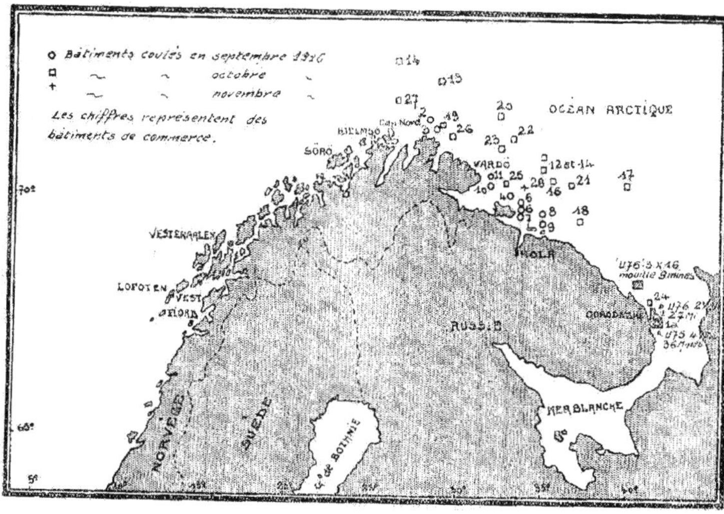


Fig 20. Arctic Ocean Operations. Ships sunk: °= September, □ = October, †= November 1916.

Meanwhile, trade warfare on the basis of Prize Rules, was in full swing according to the Kaiser's order of October 6.

"U20" led by Schwieger departed Emden on October 13 to the west coast of England. On October 18 off Cape Wrath at the entrance to the Minch he sank a British steamer, on October 23 another English and one Italian steamer, as well as a French topsail schooner (total 8,824 tons). Schwieger was well on his way back with his tried-and-tested boat when on the evening of November 2, heading from the Shetlands to Norway, he heard distress signals from "U30", which had a diesel engine out of order and the second could only generate half speed. Then on November 3 at 2 a.m., "U20" stood north of Utsire together with "U30" and stayed with this boat to provide assistance. Since both of them were making very slow progress, navigation became very imprecise, and when approaching Bovbjerg on November 4th, both boats were stranded in the fog. "U30" was freed after a short time; "U20" on the other hand was driven over the bar and stuck helplessly affixed. All towing attempts made by the IV Torpedoboat Half-Flotilla the next day proved to be unsuccessful; "U20" had to be abandoned and blown up around noon on November 5th. The entire crew was saved. Thus ended the boat's most successful war career, 7 sailing vessels and 30 steamers, including many large and very large, had fallen victim to its weapons. Failure haunted "U30" from the very beginning: On June 22, 1915, the boat had sunk at Borkum-Reede during a diving exercise. It was raised and repairs had taken so long the boat was only able to leave for its first operation under the leadership of the KptLt. Grünert on October 24, 1916. Right on its first trip it became the indirect cause of the loss of another boat. Before that, it had managed to sink a Norwegian and an English steamer totaling 1,945 tons. "U30's" accident caused even more misfortune.(diving mishap referenced earlier) Two ships from the 3rd Battleship Squadron were hit by an enemy submarine, each with a torpedo ("Kronprinz" and "Grosser Kurfurst"), which had sortied to cover the towing attempts, resulted in the loss of these ships for several weeks.

"U24" under KptLt. Walter Remy was in the northern North Sea from October 20 to November 5. On October 27, the English auxiliary cruiser "Teutonic" passed in front of its torpedo tubes between Iceland and the Faroe Islands; it narrowly escaped sinking. After sinking a steamer with contraband 75 nm west of Stavanger, the boat entered the Ems again on November 5. (see Fig 20.)

"U32" under the leadership of KptLt. Hartwig set out for the Mediterranean on October 16. On October 18 the Norwegian steamer "Athene" (1,846 tons) was sunk in the North Sea. Further successes followed in the Mediterranean (see the Mediterranean section).

"U43" after a successful Arctic Ocean cruise was away from operations in this period due to protracted repairs, returning only on January 11, 1917 to cruise, particularly to the English west coast and Spanish coast. On January 23, the Norwegian steamer "Donstad" was captured as a prize and from then on served as a cartel for the crews of sunken steamers. On January 29, he landed the crews of various sunken steamers in Corcubion Bay in Spain south of Cape Finisterre. The Italian steamer "Famiglia" later performed similar services until it was sunk. In total, 4 steamers and 3 full rigged ships were sunk during the operation with a total of 24,590 tons. On February 13, when unrestricted submarine warfare had already begun, the boat returned to Helgoland. This cruise offers a prime example of how the trade war according to Prize Rules could be waged at the time. It was still possible to use a captured steamer to accommodate steamer crews; of course the operations had to be carried out at a greater distance from the bases of enemy antisubmarine forces. The Spanish northwest corner to Cape Finisterre was particularly suitable for this. Although "U43" was the first and least successful boat in the "U43" to "U50" series, it achieved a total sunk tonnage of around 120,000 tons.

"U44" - Wagenführer - sank the English steamer "Thurso", 1,244 Br.R.T., 10 nm northeast of Peterhead on a cruise from September 21 to 29, 1916. For the last time in this war period, "U44" cruised from January 1 to January 25, 1917 to the west coast of Great Britain. The boat sank the English steamer "Baron Sempill" (1,607 tons) 120 nm SSE of Fastnet and brought three English fishing trawlers in from the North Sea. Results were affected by a particularly large number of engine malfunctions (the port screw shaft broke down on January 9).

"U46" under KptLt. Saalwächter - the actual commander, KptLt. Hillebrand, was ill - carried out an operation from December 9, 1916 to May 5, 1917 on the west coast of Ireland, the north and west coast of Spain. Although it was Saalwächter's first cruise, it brought a total of 15,785 tons. 7 steamers, 1 sailing vessel, 1 fishing trawler were brouht down; the Danish, Japanese, English, Norwegian, Spanish and French flags were among the sunk. The boat operated mainly at Ushant, Cape Finisterre, Cape Vilano and on the Spanish north coast.

"U48", KptLt. Buß, departed on December 27, 1916 to conduct trade warfare on the west coast of Ireland. Just at that time a new English minefield had just been planted in the German Bight. Mine destruction ships sweeping the night before provided “U48” with a specific route. There was much concern since no message was received from "U48" indicating safe passage through the mine area; "U48" reported its location on New Year's Eve 1916 after passing through Fair Island and sinking the Russian steamer "Tuskar", 3,443 tons, which on the way from Arkhangelsk to London.

Thus the year 1916 ended for the III Submarine Flotilla without any losses due to enemy action and with considerable sinking results.[151] Further activities of "U48" was particularly successful; sinking ten more steamers and one sailing vessel resulting in 32,864 tons, including the French steamer "La Ville du Havre" – 5,026 tons - 200 nm north of Cape Vilano on January 6 and the English steamer "Windsor"[152], 6,055 tons, at the southern tip of Ireland on January 19. The hitherto neglected Bay of Biscay had proven to be an extremely successful area for operations. The conduct of trade warfare according to Prize Rules in this most distant area from the enemy coast did not encounter any difficulties. The arrival of the KptLt. Buß in Helgoland on January 26, 1917 was accompanied by a great triumph. (The Ems submarines had to be moved from the Ems to Helgoland in mid-January because all estuaries experienced very strong icing.)

"U49" under KptLt. Hartmann set sail from Borkum-Reede on October 23. It was the first undertaking in the trade war for this boat, which only became operational in the North Sea front on August 6th. Orders were given to operate on the south-west coast of England – Ireland and south to the Spanish coast. This operation was characteristic in more ways than one. It was one of the first enterprises of trade warfare under prize regulations and the commander's first major foray. Nonetheless, it ended with a result of approximately 36,000 tons, 10 steamers, 3 fishing trawlers, 1

---

[151] Footnote in original. I wish to commend here my flag lieutenant, Oblt. Rhein, with special gratitude. He always gave me excellent support when drafting operational orders and ingeniously foresaw the preventive measures required to counter the presumed hostile defense. (The Author)

[152] "Windsor" had been previously sunk. The vessel identified here should be steamer "Nailsea Court" of 3,295 tons.

sailing vessel - including the American steamer "Columbian", 8,580 tons, with contraband - the English steamer "Lady Carrington", 3,269 tons. Since the boat had orders to stop steamers only at distances over 4000 m, the ensuing artillery battles were harmless when stopping an English steamer on November 11 and another steamer on November 12.

A very similar first success was "U50"- Commandant KptLt. Berger. The boat left Borkum-Reede on October 27, worked mainly in the Bay of Biscay, sank 8 steamers, 1 fishing trawler and took the Norwegian steamer "Older" as a prize. He lost sight of it in a storm on November 16 and did not re-establish contact.[153] Among the other steamers were the English "Bogota" – 4,577 tons - and "Sarah Radcliffe" – 3,333 tons -. The English steamer "Morazan" – 3,486 tons - was only sunk after a violent artillery battle. Also among the sunk were the Greek steamers "Styliani Bebis" – 3,603 tons - and "Ioannis" – 3,828 tons - with contraband. They were surprised in the Bay of Biscay. Upon entering the Ems on November 23, KptLt. Berger could look back on a good initial success (25,310 tons).

According to a special order from the Navak Staff, "U52" - KptLt. Walther (Hans) - and "U47" – KptLt. Metzger departed Helgoland on November 15, 1916. – to conduct trade warfare off the Canary Islands. Both boats entered Cattaro - "U52" on December 24th, "U47" on December 27th. During this trip, "U52" east of the Berlenga Islands hit the French battleship "Suffren" without any escort and sank it with a torpedo. The ship was lost with all hands. The sinking was only known due to radio communication from "U52". For the rest, the boats in the Canary Islands only managed to sink the Greek steamers "Salamis" – 3,640 tons - and "Spyros" – 3,500 tons - ("U47") and the Dutch steamer "Kediri" – 3,784 tons - as well as the French sailing vessel "Emma Laurans "("U52"). This area of operations was not particularly rich in merchant ships. "U47" had already sunk a Dutch and a Norwegian sailing vessel in the North Sea 145 nm northwest of Horns Reef and sank the

[153] "Older" was recaptured by the auxiliary cruiser HMS "Otway" on November 13, 1916.

Norwegian steamer "Leo"[154] – 1,823 tons - 20 nm west of the Alboran Island. "U47's" M.A.N. engines had been so strained by this trip that the boat did not return from Pola until May 1, 1917.

"U54" - KptLt. Frhr. v. Bothmer - posted to the Murman coast in mid and late October, was unable to achieve any success there due to the great caution shown by the steamers, the earlier efforts of "U43", "U46", "U48" had made an impact.

"U55" - KptLt. W. Werner - left Helgoland on October 14 in order to meet "U53" - KptLt. Rose -on his return from America (see pp.200-201). "U53" was actually sighted on October 23rd, but needed no assistance. The operation was used by "U55" to sink the English steamer "Marina" – 5,204 tons - 26 nm WNW from Skellig-Island.

The name of another English steamer of around 5,000 tons sunk on the 24th of October from Flannan-Island could not be established.[155]

"U56" - KptLt. Lorenz - departed Helgoland on October 13 to conduct trade warfare on the Murman coast. It did not return from this operation; on November 2, 1916 it was sunk in the Northern Arctic Ocean in artillery battle with two armed steamers.[156] One can see the importance England placed on communications with Russia on the Murman coast that strong security was immediately organized after the initial successful submarine operation there.

"U57" - KptLt. Georg - conducted trade warfare on the west coast of Ireland from October 13th to November 5th. On October 18, Georg had an artillery battle with two fishing trawlers. On October 23, the English sloop "Genista" – 1,250 tons – was sunk by two torpedo hits while escorting a steamer. The main success, however, was the sinking of the English steamer "Rowanmore" – 10,320 tons - 128

---

154 This vessel's name is recoreded as "Sno" to p.331, vol. 3 of the German Official History, Handleskrieg mit U-booten.

155 British steamer "Clearfield". p.249, Ibid.

nm WNW from Fastnet Rock, a fishing trawler and the Norwegian steamer "Saturn" – 1,013 tons - were also destroyed.

Even more successful was "U57's" venture from January 9 to February 7, 1917. Among other vessels, sank the English steamer "Manchester Inventory" – 4,247 tons - 50 nm northwest Fastnet Rock, the English steamer "Trevean" – 3,081 tons - 240 nm south of Fastnet Rock, the Portuguese sailing vessel "Minho" in the same area and the Italian steamer "Bisagno" 100 nm northwest of Clifden.

"U57" was a boat that suffered from frequent, often severe machinery malfunctions. Nevertheless, its commander, KptLt. Georg, knew how to make the boat a great success, especially when working close to the enemy coast.

"U58", KptLt. Wippern, which joined the North Sea submarines in October, worked from October 24th to November 2nd between Skagerrak and Peterhead and from November 22nd to December 11th scouted for the auxiliary cruiser "Möwe" when it departed on its first venture in the Atlantic Ocean. A steamer and two barques were sunk from these trips (1,436 tons).

"U59", KptLt. Frhr. v. Fircks, entered the operations area of the Commander of Submarines on November 20, sunk a Swedish three-mast and a two-mast schooner with contraband goods (407 tons) on his first North Sea outpost cruise to Horns Reef. The first operation into the Atlantic Ocean in the area encompassing Cape Vilano, Cape Ortegal and Ushant resulted in 7,662 tons. The largest sunken steamer was "Solvang" (Norwegian) of 2,970 tons.

"U63" under KptLt. Otto Schultze, who had distinguished himself in the submarine operation associated with the Skagerrak battle and sank the English small cruiser "Falmouth" on August 20, 1916, left Borkum-Reede on October 16, 1916 for his crossing to the

---

[156] "U56" Survived this convoy attack/counterattack battling the Russian old destroyer "Grozovi", dispatch vessel "Kupava" and British minesweeper "T-13". It is presumed the boat succumbed to damage afterwards.

Mediterranean. By the time it entered Cattaro on November 6, 24,516 tons had been sunk, including the English flag three times, Norwegian twice, Italian twice, American once, and Greek once; 7 sinkings occurred before passing the Gibraltar Strait, 2 after the passage. The further cruises of "U63" are reserved for Mediterranean Sea operation descriptions.

"U64" under KptLt. Morath was also no longer operating against trade in the North Sea; on October 26, this boat was also dispatched to the Mediterranean. On November 10, a Norwegian and a Danish steamer were sunk (a total of 6,801 tons). On November 13, the boat had an artillery battle off the coast of Algiers with the armed English steamer "Islandia" (2,069 tons), which had to be broken off. On November 15, the English steamer "Matarazzo" (2,823 tons) was sunk with a three torpedo salvo. On November 19, "U64" arrived in Cattaro.

"U65" under KptLt. von Fischel also went to the Mediterranean on October 26 and arrived in Cattaro on November 18. Due to fuel oil shortage and delays to bad weather, the boat did not operate miltiarily during the crossing.

"U63", "U64" and "U65" were particularly high-quality submarines built at the Germania shipyard in Kiel with M.A.N. engines originally intended for Russian submarines. The boats easily ran to 16 knots. Every commander, every flotilla was envied for their possession.

"U66", Kptlt. v. Bothmer, was planned to be on the west coast of England from October 13 to November 3. On October 30, the cruise had to be stopped due to persistent stormy weather. On the return voyage v. Bothmer sighted six English battlecruisers in the North Sea on November 1, which he had occasionally sighted during fleet operations on August 19. Unfortunately, the two torpedoes fired missed their target because the battlecruisers' high speed did not enable the boat to get close enough.

On the afternoon of November 30th v. Bothmer departed again from Kiel, together with the auxiliary cruiser "Wolf" - FKpt. Nerger - to escort the cruiser during the initial voyage, which succeeded, because "Wolf" was able to break through the English blockade and reach the Atlantic Ocean without hindrance. "U66" had "Wolf" in sight to December 5, and still in radio communication until December 8. On this cruise, the Norwegian steamer "Bjor" —1,190 tons - and the Swedish three-mast schooner “Palander” - 310 tons - were sunk with contraband off the Skagerrak and Kristiansand respectively. On December 16 v. Bothmer returned to Kiel. An English captain was taken from an examined Swedish steamer and brought with him.

"U67" - KptLt. Nieland - was enroute to Vestfjord from November 21st to December 18th with orders to stop Arkhangelsk traffic. It turned out that almost all steamers were heavily armed. Nieland was therefore unable to achieve a success because torpedoing without warning was prohibited.

From January 13 to February 25, Nieland was again on the English west coast. The cruise thus extended into the unrestricted submarine war phase. On January 28, 60 nm from Cape Ortegal the Danish steamer "Daisy" (1,227 tons), on January 29 the Spanish steamer "Punta Teno" (1,042 tons) 25 nm from Cape Ortegal, on February 1 the Spanish Steamer "Butron" (2,433 tons) off Bilbao, on February 2 the Greek steamer "Eidikon" (1,166 tons) 70 nm east of Cape Ortegal, on February 5 the Peruvian barque "Lorton" (1,419 tons) 55 nm sunk west of Bilbao. On February 19 near Eddystone Light torpedoing a steamer of about 4,000 tons which was salvaged.

"U68" - KptLt. Güntzel - had not returned from English the west coast from an operation after departing from Emden on March 15, 1916; the boat had been destroyed by the submarine trap "Farnborough" near the Scilly Islands. (See page 140.)

"U69" - Kptlt. Wilhelms - cruised from October 13th to November 4th to the English west coast and the Scilly Isles. The overall result of the undertaking, carried out with great enterprise, was 17,236 tons, including the English steamer "Cabotia" (4,309 tons) 120 nm from Tory Island, which fought hard so that it had to be beaten down with artillery and the captain and 31 crew killed. In the area about 70 nm west of the Isles of Scilly, the following steamers fell victim to Wilhelms from October 24 to 26: the Norwegian "Sola" – 3,261 tons -, the English "Rappahanock" – 3,871 tons - and the English "North Wales". On the return trip, 95 nm from Hellisö (Norway) the English steamer "Spero" – 1,132 tons - and in front of the Skagerrak entrance the Swedish barque "Bertha" - 591 tons - were sunk.

KptLt. Wünsche, commanding officer of "U25" since the beginning of the war, now commanding officer of "U70" since early 1916 carried out a successful cruise from December 15, 1916 to January 12, 1917. He drove through the English Channel and on December 17, sank 10 nm NNE from Cape de la Hague the English steamer "Pascal" - 5,587 tons – with a torpedo and took the captain prisoner. Afterwards the French sailing vessel "Eugene Gaston" - 183 tons - was sunk on December 18, 30 nm from Ushant, on the same day it was the turn of the English steamer "Flimston" – 5,751 tons - and the French sailing vessel "Hirondelle" - 148 tons; both were destroyed by explosive charges. The captain and chief engineer from the "Flimston" were taken prisoner. In addition, Wünsches cruise took him to the Spanish coast and on the return voyage to the north of England; a total result of 16 vessels of 26,183 tons was sunk; the Italian, English, Norwegian, Greek, French and Russian flags were represented among the sunken ships. A particularly valuable ship was the Greek "Tsiropinas", sunk 105 nm north of Cape Ortegal on New Year's Day 1917.[157] The attempt to direct the Norwegian steamer "Thyra" - 750 tons - from Cape Ortegal to a German port failed; the ship was intercepted by English patrols and returned to the Norwegians.

"U71", although a mine boat, under KptLt. Hugo Schmidt was also used for trade warfare. From October 14th to November 4th 1916 the boat was in the Skagerrak. Two Swedish and one Norwegian

---

[157] "Tsiropinas", on a voyage from Buenos Aires to Belfast with a cargo of maize.

steamer and a Swedish sailing vessel, totaling 4,544 tons, were sunk. Cargo was thrown overboard on the Danish steamer "Eos". On a second cruise from December 11th to 23rd on the southern coast of Norway and the north coast of Scotland, two Norwegian steamers, a Dutch and a Danish sailing vessel, were sunk – 1,890 tons. Another cruise to Skagerrak and the North Sea only led to the investigation and release of several neutral steamers. The operation had to be canceled on January 19 due to bad weather and damage to the starboard engine. On January 21, the boat returned to Helgoland.

"U75" - Beitzen -, whose mine success against "Hampshire" - Lord Kitchener - was mentioned in the first part of this book, was in the Skagerrak from November 2 to 30. The low engine power of these large mine boats did not make their use in the trade war on the west coast of Great Britain seems appropriate. On November 22, the Swedish steamer "Reserv" was sent as a prize to Kiel, on November 16 the Danish sailing vessel "Fenja" - 433 tons - 15 nm south of Kristiansand, and on November 23 the Swedish steamer "Arthur" – 1,435 tons – were sunk 10 nm east of Skagen.

Occasionally the large mine boats were utilized for laying mines in the submarine trade warfare operational areas in order to promote the former. Thus "U76" under KptLt. Bender received orders to lay mines on the west coast of Shetland between Christmas and New Year 1916. Unfortunately, the boat was forced to abandon the operation on December 26th due to an accident in the North Sea. On the evening of December 27th, it returned to Helgoland.

An operation by "U78 - Dröscher -, which was supposed to disrupt trade to Arkhangelsk, led to boarding the Norwegian steamer "Atle Jarl" on October 21 and then the temporary return of the boat to Helgoland; on October 29, it departed again and reached Trondheim Fjord. On November 16, the Norwegian steamer "Vega" – 1,204 tons - was sunk off Stavanger. On afternoon of November 22, "U78" returned to Helgoland.

"U79" (KptLt. Jeß) made on his own suggestion a particularly remarkable undertaking for a large mine boat. He drove through the Dover Straits to conduct trade warfare in the Bay of Biscay and Irish west coast. The Channel passage succeeded despite the boat's slow speed and poor handling. They evaded destroyers and patrol boats by diving in time. The cruise went to Cape St. Vincent and the Spanish coast. The astonishing result of 16,497 tons was achieved by this boat type. (8 steamers and 2 sailing vessels, including the Danish flag twice, English twice, Norwegian three times, Portuguese, Italian, Greek and Japanese each represented once.[158]) The return trip carried out around the Shetlands Islands was particularly adventurous. On January 24, Jeß stopped the Norwegian steamer "Nama" in the northern North Sea and took it with him as a prize. Since Jeß had run out of fuel on the long voyage he let himself be towed by the prize to the German Bight until he was sighted by the German destroyer "B97" on January 27 and tugboat "Voßlap" on the next day then towed by the latter to Helgoland.

Given the unfavorable characteristics of the boat, this undertaking meant a very special achievement.

A pure mining operation was carried out by "U80" (von Glasenapp) during this war period. The voyage departed to the Irish Sea on October 9, where the boat successfully escaped an enemy patrol vessel's attempt to ram it on October 16. According to instructions the mines were to be planted between the Isle of Man and Holy Head on October 27th. The English steamer "Skerries" (4,278 tons) ran into them on November 4. 15 nm NNW of the "Skerries" at Anglesea on December 18 the English steamer "Opal" (599 tons) near the Isle of Man, December 19 the steamer "Liverpool" (686 tons) ibid.[159] The boat returned to Helgoland on the morning of November 11th.

---

[158] The Greek steamer was eventually disallowed.

[159] Footnote in original. It is not clear whether the sinking of the auxiliary cruiser "Laurentic" (14,892 tons, sunk January 2, 1917 near Iceland) can be attributed to these mines or to those laid by "U79" on October 4 in front of the Firth of Clyde.

The initial activity of these large mine boats was difficult in comparison to the nimble new 800 ton Type M boats, which are now detailed for the first instance in the following text. These were the new boats, "U81" to "U86" from the Germania shipyard, whose commissioning and first operations occurred in this war period. They were assigned to the IV Submarine Flotilla on the Ems (Commander KptLt. Prause). The IV Submarine Flotilla thus received an increase in combat strength, and since these new boats were led by excellent commanders, made them far superior to any other flotilla for the time being.

"U81" came under KptLt. Weisbach, when as part of the III Submarine Half-Flotilla, transported Sir Roger Casement to Tralee Bay in Ireland in April 1916 (see pp.149-150 in Special Operations"), arrived at the front in October 1916. After a few minor cruises into the North Sea, Weisbach was ordered to escort the auxiliary cruiser "Wolf" along the Norwegian coast through the North Sea with "U83" - Hoppe. To this end, the two boats left the Ems on November 30th. Like "U66", they proceeded out of sight at a greater distance ahead of "Wolf", warning in time of suspicious smoke clouds, etc. "Wolf" under its brave commander, Kapitän Nerger, got through safely through and achieved great sinking results in his one-year undertaking. "U81" sank a Swedish and a Norwegian steamer not far from Horns Reef – 2,577 tons - and entered Helgoland on December 20.

KptLt. Hans Adam joined the IV Submarine Flotilla on November 20, 1916 with "U82". He was formerly - see page 53 - the commendable head of Special Naval Commando Pola and there he had organized the first B and C boats into the Mediterranean Submarine Flotilla. From December 4 to 11, he waged trade warfare on the Lindesnaes-Peterhead line and off the Skagerrak. Four Danish sailing vessels and the Norwegian steamer "Ella" – altogether 1,957 tons – was the haul. From December 20, 1916 to January 18, 1917, Adam then made his first cruise to the Irish west coast and the western Channel exit. The most significant sinking was the French steamer "Omnium" – 8,719 tons - on January 2; 2

Danish, 1 Italian and 1 English steamer were also destroyed – 14,153 tons -; the Dutch steamer "Gooiland" (January 4) had its cargo thrown overboard.

KptLt. Hoppe who used to be with "U22" in the III Submarine Flotilla and had distinguished himself by sinking the auxiliary cruiser "India" in front of the Vestfjord, returned to the North Sea war theater with "U83" at the end of November 1916. He was only supposed to carry out an escort for the auxiliary cruiser "Wolf" together with "U81", while doing so he also sank the Swedish three-masted schooner "Niord" 35 nm SzW from Lindesnaes; he did not return from the next operation which was during the unrestricted submarine war phase, his destruction will be reported in the fifth volume.[160]

Finally, on December 3, 1916, "U84", commanded by KptLt. Walter Roehr arrived at Borkum-Reede. He took on the new war tasks with particular vigor. From his first shakedown cruise to the North Sea, he brought in as a prize the Norwegian steamer "Aamot" on December 14 and the Swedish steamer "Malcolm" on December 18; the latter was sent to Lister Deep.

On January 1, 1917, "U84" left the Ems for the first trip to the main operating area west of Great Britain together with "U44". The steamer "Alexandrian", although attacked and damaged 20 nm SW of Fastnet on January 9 with gunfire and a torpedo was able to put into Berehaven, but the following ships were sunk: On January 10th Norwegian steamer "Bergenhus" – 3,606 tons -, 165 nm SSW from Fastnet. On January 12, English steamer "Auchencrag" – 3,916 tons - 20 nm west of Ushant. 31 crew were taken on board from this steamer who were transferred the next day to the Norwegian steamer "Morild I". The steamer "Morild I" had to throw two thirds of its cargo overboard because it was contraband. Afterwards on January 15 sank the Danish steamer "Omsk" – 1,574 tons - 115 nm south of Fastnet and on the same day the English full rigged ship

---

160 Gayer did not publish the fifth volume as mentioned in the Translator's Forward. "U83" was sunk by the submarine trap HMS "Farnborough" on February 17, 1917. Hoppe did not survive.

"Kinpurney" – 1,944 tons - 110 nm west of Bishop-Rock. The rescued crews were given to "Morild I", this ship had to be released due to a lack of drinking water. On January 20, 40 nm WNW to W of Fastnet, the English steamers "Bulgarian" (2,515 tons) and "Neuquen" (4,191 tons) were sunk. Unfortunately in the heavy weather relatively large losses of the steamer crew could not be avoided. The captain and 3 men died from "Bulgarian", the captain and 17 men from "Newquen"; Roehr took 10 men from "Bulgarian" on board. From January 21st to 25th the return voyage north of Great Britain was carried out. The total result was 17,138 tons. The operations described in some detail are typical of those carried out prior to the unrestricted submarine war phase.

"U85" under KptLt. Petz arrived at the Ems in mid-January. From January 22 to January 27, this boat made its first cruise to the North Sea. On January 26, the Norwegian steamer "Dicase" (928 tons) was sunk 40 nm southwest of Skudesnaes in accordance with the Prize Rules. Due to bad weather and fauly lifeboats, the crew of the steamer was taken on board and later handed over to a Danish sailing vessel. On January 27, Petz received the radio messages of the disabled "U79" (see page 214) and immediately rushed to the aid of this boat.

It now remains to summarize the activities of the small B and C boats subordinate to the Commander of Submarines North Sea in the II (Helgoland) and IV (Brunsbüttel) Submarine Flotillas.

"UB21", Oblt. Hashagen, was positioned from October 1 to 11, 1916 on the northwest side of the Dogger Bank for trade warfare. 2 Swedish, 1 Norwegian sailing vessels and 1 Norwegian steamer were sunk, 1 Norwegian steamer, 1 Danish three-masted schooner and 1 Norwegian motor ship were brought in as prizes (2,009 tons). From December 4 to 13, 1916, "UB21" was commanded by Oblt. Walther to stalk enemy submarines in the area of Horns Reef, and from January 16 to 23, 1917 on the coast of England However, due to the heavy seas, no steamers could be stopped.

Oblt. Putzier's first trade war operation was quite successful, "UB22", went to the east coast of England from October 18 to November 5, 1916. A total of 4,551 tons were sunk; 2 Norwegian steamers, 3 Norwegian sailing vessels, 1 Swedish barque and 1 small Danish steamer. Cargo was destroyed from the Norwegian schooner "Alliance" 90 nm east of Fern Island, and the Norwegian steamer "Pluto" was brought in as a prize.[161] The further cruises by this boat, until the beginning of unrestricted submarine warfare, were intended to hunt warships.

"UB27" under its proven commander Oblt. Dieckmann, cruising from October 2 to 14, 1916 in front of the Firth of Forth, torpedoed the English steamer "Jupiter", which could still be towed to safety, and sank the English fish steamer " Magnus "(154 tons) 40 nm ENE from Longstone. Under its new commander Oblt. Lübbe, the boat carried out three patrols into the North Sea until the beginning of unrestricted submarine warfwar without result. On last voyage [ed.- in this period], the boat had two battles with submarine traps on the English east coast which dropped numerous depth charges.

"UB34" under Oblt. Theodor Schultz managed to sink 9 vessels (totaling 5,466 tons) at Peterhead and on Doggerbank in two trips from October 18 to November 3 and from December 7 to 23. A third trip in January 1917 was unsuccessful because of the same counteraction as "UB27" had found. (Submarine trap)

"UB35" had a particularly successful voyage from October 13 to November 2, when Oblt.v. Schräder had been assigned to command the boat from the KptLt. Gebeschus. On the Scottish coast north of Fern Island, 4 Norwegian steamers, 2 Norwegian sailing vessels, 1 Danish steamer and 1 Danish schooner were sunk (totaling 5,826 tons). Oblt. v. Schräder was later promoted for this excellent achievement.

---

[161] Presumably a draft compilation error in the original; p.274, vol. 3 of the German Official History, Handleskrieg mit U-booten indicates "Pluto" was captured by "UB21" at this time.

The last B-boat to be mentioned in the North Sea war theater in this period - "UB41" - under the command of Oblt. v. Sichardt was only able to bring in the Norwegian motor-schooner "Thyholm"[162] as a prize on November 21 to List on the evening of the 31st. The English fishing trawler "Cetus", damaged by artillery on January 18, was safely brought to Blyth in tow.

It remains to describe the activity of the improved mine submarines (C II boats), which in late 1916 arrived in the command area of the Commander North Sea Submarines to the I Submarine Flotilla (Brunsbüttel). These boats were a particularly successful type, as they were superior to the improved small torpedo submarines (B II boats) in terms of surface speed by 2 1/2 to 3 knots and, in addition to 14 or 18 mines, held two 50 cm bow torpedo tubes and a 50 cm stern tube. Their range of action was 8000 to 10000 nm.

“UC29 "under Oblt. Rosenow, who on" UC24" received his training from the tried and tested KptLt. Schneider, planted mines in front of the Firth of Tay on November 2. From November 22nd to 29th the boat was enroute to escort and provide security for the auxiliary cruiser "Möwe", but had to turn around on the evening of the 26th off the Norwegian coast due to a mechanical breakdown. In the second half of January Rosenow was again in the North Sea. On January 21, two mine fields were planted in front of the Firth of Forth, on January 22, mines were laid in front of the mouth of the Tay. The English steamer "Clan Shaw" (3,943 tons) ran onto these on January 23. The return voyage commenced after mines were laid before Aberdeen on January 23. On January 24, Rosenow stopped the Norwegian steamer "Haakon VII", which had 50 passengers from an enemy nation on board. 6 Russian engineers and mechanics were taken off. After the Norwegian steamer "Sunniva" (589 tons) was sunk on January 24, "UC29" arrived in Helgoland on January 27 (result: 4,532 tons).

162 This vessel’s name is recoreded as “Thyholmen” in other registries.

"UC31" under Oblt. v. Schräder planted mines in the evening of December 30th before Tynemouth and before Sunderland. The pilot cutter "Protector" (200 tons) ran onto this on December 31, and also the English steamer "Lonclara" (1,294 tons) on January 4. 19 crew and the captain were killed.

On January 27th Schräder again in front of Sunderland laid five minefields. The English steamer "Ravensbourne" – 1,226 tons - ran on them on January 31. The Danish steamer "O. B. Suhr "(1,482 tons) was sunk 150 nm WNW from Horns Reef. Furthermore, 3 English fishing trawlers fell victim to v. Schräder (2,001 tons total). On January 29 v. Schräder was prevented by the intervention of an English patrol ship from sinking further English fishing trawlers 130 nm east of Farn Island.

"UC32" under Oblt. Breyer planted mines on December 13 near Newcastle and on December 14 off Sunderland. On the same day the English steamer "Burnhope" – 1,941 tons - fell victim to these. On January 31, 1917 Breyer again laid his mines in three fields in Tees Bay, on which the tug "Ida Duncan" - 139 tons - went down on the same day. On the same cruise he sank the Swedish steamer "Edda" - 535 tons -. It was the last cruise for "UC32". The boat did not return from its first cruise during the unrestricted submarine war; it was destroyed by a mine explosion off Sunderland. Breyer and part of the crew were saved.

"UC33" under KptLt. Schelle was at sea from December 31, 1916 to January 7, 1917, minelaying the entrances to Lerwick (Shetlands), Kirkwall (Orkneys) and the Firth of Moray. On January 2, six mines were planted after observing the entry route at Kirkabister (Shetlands); on January 3, two fields in Skapinsay Sound (Orkneys), on January 4, two fields in the Firth of Moray. There is no apparent success in these mines.[163] On January 7th the boat returned to Helgoland.

---

[163] Footnote in original. The English trawler "Cusion" (209 tons) sank there on February 12, 1917.

"UC34" commanded by of Oblt. Sprenger departed on December 18, 1916 from Helgoland to the Mediterranean Sea, where this boat was later achieved great success. On December 27, Sprenger sank the French sailing vessel "Maud" and on February 28 the Russian sailing vessel "Seedonis"- a total of 460 tons -.

"UC35" under Kptlt. Ernst von. Voigt departed on December 3 from the Elbe to Zeebrugge and from there on the evening of the 7th through the Channel to the Mediterranean.

"UC37", Oblt. Launburg, also went through the English Channel to the Mediterranean on December 23. We will encounter these boats again when describing Mediterranean activity.

"UC40" under Oblt. Deuerlich carried out two short North Sea operations at the end of December and was then sent on a mining operation to the English east coast from January 13 to 23, 1917. On 16 and 17 January when approaching the English east coast near Blyth, the boat encountered strong patrols, twice involuntarily grounding. On the 16th and 17th six mines were laid in the Blyth roadstead, four mines at Coquet Island, four mines at Sea Home, and the same at Longstone. The mines were planted after previous observation and with good land bearings. The English steamer "Hildawell" – 2,494 tons - probably fell victim to them on December 20, 1916, including the entire crew of 22 men.[164] An attack approach on an English armored cruiser was carried out twice on 19 January, but the boat did not fire due to the long distance.

"UC41", KptLt. Bernis, was from December 30, 1916 to January 10, 1917 minelaying on the English coast (Bermick, Aberdeen, Firth of Forth, Tay estuary).[165]

---

[164] pp.281-2, vol. 3 of the German Official History, Handleskrieg mit U-booten credits this ship to minefield No. 17 laid by “UC32”.

[165] Footnote in original. On February 9, 1917, the English drifter "Desto" (229 tons) sank in front of Aberdeen.

"UC43", KptLt. Sebelin, brought in the Norwegian steamer "Lupus" as a prize on December 31 from the North Sea. On January 11th and 12th the boat placed mine fields off Whitby with such careful observation of the steamer routes that on January 12th the English steamer "Brentwood" – 1,192 tons – 4 nm ENE from Whitby ran onto them and sank. On January 13, mines were laid again at Hartlepool.[166] On January 17, the boat returned to Helgoland due to damage to the Bamag coupling.[167] It was the last undertaking of "UC43"; on the next voyage it was torpedoed on March 18, 1917 by the English submarine "G13" north of Mückle Flugga Light. (Shetland Islands). English submarines lurked for ours at navigationally distinctive control points. Experienced boats used to pass such places during the day either with high speed and zigzag courses or submerged.

We have reached the end of submarine activity in the North Sea (operations area of the Commander of Submarines Kapitän Bauer[168]) in this war period. It may come as a surprise that the work of the B II and C II boats has been described in such a relatively detailed manner, while later the more successful operations in the Mediterranean are dealt with much more summarily. The reason is that these small ventures in the North Sea are particularly characteristic of operations. Often the submarine front was believed this infinitely arduous North Sea submarine activity could have been supported to a much greater extent by operations by fast surface strike forces, which incidentally, Supreme Command Order O 31218 I issued on October 6, 1916 also applied to the Fleet and Marine Corps.[169] This command contained the passage: "In the North Sea trade war, if possible, by surface forces." With such "raids", many Scandinavian steamers could have been sunk with contraband in one day as was done by submarine activity in weeks,

---

166 Footnote in original. According to English information, the destroyer "Pheasant" (1,025 tons) ran onto a floating mine off the Orkneys on March 1; a connection could be possible.

167 Footnote in original. Between the Diesel engine and double dynamo.

168 A typo as Bauer was promoted to Kapitän zur See in October 1917.

169 Flanders naval assets operated under the aegis of Marine Corp Flanders which also included a land and air component.

especially since we had the opportunity to undertake such surface operations not only from the German Bight but also from Skagerrak to the North Sea. This would have given us the opportunity to send even more of the new B and C boats into the Mediterranean because considerably larger sinking results were possible there and we ultimately waged the trade war against "world tonnage".

## Flanders in the Period from October 1916 to the end of January 1917.

### Flanders January 1917.

The general order for commencing trade warfare according to Prize Rules mentioned on page 222 (Supreme Command O 31218 I of October 6, 1916 to the Fleet and Marine Corps) contained the following important instructions for Flanders submarine activity: There is no need to check the papers of merchant ships if his appearance or other behavior reveals his hostile nationality. By Supreme Command order dated October 10, 1916, the area south of latitude 56° and east of the longitude 7° was established for mine warfare by Marine Corps submarines. According to a Supreme Command order dated October 17, torpedoes could be used immediately against ships that opened fire when stopped for inspection, with the exception of passenger ships. According to a Supreme Command order of October 24th, tankers, on which armament was correctly recognized, were allowed to be sunk without warning. A Supreme Command order of November 3rd said: Ships anchored in the roadsteads of Cherbourg, Le Havre, Boulogne, Calais within territorial waters may be sunk without warning; Passenger steamers and hospital ships were exempt, no ship could be sunk in Dieppe's roadstead without warning; a further order of November 4, 1916: In the Channel, east of the Cape de la Hague — St. Albans Head line can be sunk without warning on the England-France route and vice versa and darkened vessels. Belgian relief and hospital ships should be allowed through.

In many cases the contents of all orders show that experienced submariners who were familiar with conducting submarine warfare had not participated in their preparation. Almost all of the orders consistently showed the Naval Staff's concern regarding involvement with America; the Naval Staff cannot be blamed for this.

Let us now consider the operations originating from Zeebrugge or Ostend:

The following new boats arrived in the theater:

| | | | | |
|---|---|---|---|---|
| on October 3. | "UB40" | under Oblt.z.See. | | Karl Neumann, |
| 2. | "UC17" | - | - | Wenninger, |
| 19. | "UC18" | - | - | Kiel, |
| November 9. | "UC19" | - | - | Nitzsche, |
| 14. | "UC21" | - | - | Saltzwedel, |
| 29. | "UC46" | - | - | Moecke, |

Again a significant increase in boats and very powerful ones.

"UB6" under Oblt. von Heydebreck, from January 10, 1917 onwards under Oblt. Steckelberg, made ten trips to outpost positions off the Flanders coast and in the Hoofden. On January 18, the boat was attacked by an enemy submarine with three torpedoes that failed; "UB6" being only a very small target. On December 8, 1916, an enemy submarine tried to ram "UB6".

"UB10" under Oblt. Amberger was utilized in the same way. On October 4th, 19th, and November 27th, Amberger shot at enemy destroyers; but without success due to the vessels' high speed. On eleven trips off the Flanders coast and in the Hoofden, numerous skirmishes with other destroyers, enemy aircraft and submarines were successfully withstood. In January 1917 the boat sailed to the same task under the commands of the Oblts. v. Rohrscheidt and Pilzecker. Also a trip to Nordhinder Lightship from February 13 to March 2, 1917 under Oblt. Graf v. Schmettow did not achieve any success due to the strong security of steamers encountered.

The following boats operated in similar trips with no notable results, "UB12" under Oblts. Gerth and Niemer, "UB16" under Oblt. Müller-Schwarz and "UB17" under Oblts. Degetau and Meier (Ulrich). The latter brought a Norwegian steamer to Zeebrugge as a prize on December 15. From a military point of view, the long journeys of these small B-boats off the coast and in the Hoofden naturally had considerable operational value as they kept enemy patrols constantly on guard. The constant and exhausting activity of the commanders of these small boats on duty off the Flanders coast must also be commemorated. On October 26, "UB16" served as a "lightship" for the III and IX Torpedoboats Flotillas push into the Channel. The advance of the IX Torpedoboat Flotilla (KKpt. Goehle) into the Dover Straits was also supported by illuminating buoy 2501 by "UB16". This sortie was supposed to roll up enemy patrols in Dover Straits to clear the way for the Flanders boats.[170]

Steinbrinck ("UB18"), whom we left (page 195) after his successful (36,125 tons) cruise from September 2 to 16, 1916, was again from October 19 to 27, 1916 in the western part of the Channel and at the Scilly Isles. This undertaking had a result of 8,537 tons, including 3 English, 3 Norwegian steamers, the French barque "Condor" of 760 tons and the big French sailing ship "Briceny"[171] – 2,197 tons - and "Cannebiere" – 2,254 tons -. On October 21, during the sinking of the "Briceny", the boat was fired upon by an auxiliary cruiser. The fire was returned and the boat submerged. On October 22nd, a steamer stopped by a warning shot raised the English battle ensign and turned out to be a formidable adversary. Steinbrinck escaped his fire in time by diving; since the vessel was apparently a submarine trap with very shallow draft, a torpedo shot was unsuccessful. In the afternoon of the same day, Steinbrinck had to endure an artillery battle with another English steamer.

Oblt. Lafrenz, who took over "UB18" after this trip, carried out a very successful operation to the western Channel after two smaller cruises into the Hoofden from November 22nd to December 5th. He sank 11 steamers and 10 sailing vessels totaling 11,903 tons. The

---

[170] Footnote in original. In the course of this sortie, six Dover Patrol guard drifters were sunk on October 27, one seriously damaged. The destroyer "Flirt" (380 tons) was also sunk. The destroyers "Zulu" (1,027 tons) and "Nubian" (1,062 tons) were badly damaged. They were later assembled as onen a boat "Zubian".

[171] The name is “Brizeux” according to p.285, vol. 3 of the German Official History, Handleskrieg mit U-booten.

Norwegian, Danish, Spanish, English and French flags were represented. On November 24, during the sinking of the Norwegian steamer "Oifjeld" – 1,998 tons - north of Cape d'Antifer, "UB18" was shot at by two French destroyers and a fishing trawler. On December 2, south of Wolf Rock, even a sailing vessel opened fire. On December 4th in the evening at Casquets Islands, "UB18" was depth charged by a destroyer; soon afterwards the boat got caught in a strong net and escaped when torn apart by deep diving. Another trip to the Scillys and Casquets from December 27, 1916 to January 9, 1917 resulted in 7,879 tons (4 Norwegian, 2 Danish steamers).

"UB19" under Oblt. Becker from October 22 to November 3 operated in the western Channel to the north of Calais. The following were sunk: the Belgian steamer "Comtesse de Flandres" – 1,410 tons - and the French steamer "Iduna" – 1,65 tons -.

On November 10, the boat went under its new commander, Oblt. Noodt, to the Hoofden and took the Dutch mail steamer "Koningin Regentes" 5 nm NW-Schouwenbank as a prize to Zeebrugge, because he was suspected of throwing packages overboard when the submarine approached.

On November 22, 1916, Noodt was on his last voyage, which led to the western Channel. On November 27, the Norwegian steamers "Belle Isle" – 1,883 tons - and "Visborg" – 1,311tons - were sunk, then on November 30 near the Scillys "UB19" fell victim to an English submarine trap.

"UB23" - Voigt - was from October 18 to 31 in the western Channel by Ushant. In addition to the Norwegian steamer "Snestad" – 2,350 tons - and the Norwegian "Venus II" - 784 tons - only smaller vessels – 3,896 tons in total - could be sunk because the boat got into artillery battles with steamers on almost every day of the operation. The extent to which the English countermeasures had developed in the Channel was also evident in the two following cruises of "UB23", under which Oblt. Ziemer led the boat. On November 30, Ziemer sank the French sailing vessel "Grete" - 170 tons -, on December 2 the English steamer "Harpalus" – 1,445 tons - 34 nm SSW from Galley-Head, on December 4 the Norwegian steamer "Nervion" – 1,921 tons - 30 nm south of Fastnet, and on December 7th the English tanker "Couch" – 5,620 tons - 12 nm SW1/2W from Anvil Point (altogether 9,156 tons), on November 28th it started a fire on a steamer, and on November 30th sank the fishing trawler "Gaete". On December 1, a stopped steamer turned out to be an auxiliary cruiser and fired at the boat, the submarine attack failed due to skillful maneuvering by the auxiliary cruiser ("Dykland", 4,291 tons).[172] On December 7th a torpedo attack on an auxiliary cruiser was unsuccessful; apparently the torpedo apparently struck the seabed and detonated. On the subsequent cruise to the western Channel from January 2 to January 15, 1917, only the English sailing vessel "Brenda" - 249 tons - was sunk 10 nm SSW by Beachy Head; on January 12, "UB23" came across the English submarine trap "Kai - Danmark".[173] When Ziemer approaches closer after a boat has been launched from the stationary "Kai", he receives a volley from four guns. Although the boat is damaged, it is still able to travel and dive and must beat a retreat homeward; after be stuck 1 1/2 hours in the Channel passage and incorrectly entering the Scheldt, UB23 successfully returns to Zeebrugge on the afternoon of the 15th.

Pustkuchen directed "UB29" on a cruise from October 20 to 30, and was just as successful as in September. Operating in the Channel as far as Wolf Rock, he sank four steamers, 2 sailing vessels and 1 cutter totaling 8,831 tons, including the Greek 3,636 ton steamship "George M. Embiricos", 18 nm SE-Lizard, and the English steamship "Sidmouth"- 4,045 tons - 22 nm south of Wolf Rock. Even though "UB29" was shot at by a sailing vessel and once by a destroyer, the extraordinarily smooth cruise progress of veteran Flanders commanders compared to new commanders is striking. The knowledge old Flanders warriors had learned in 0-3 years could

[172] "Dykland" was actually an armed merchant steamer not a naval vessel.

[173] "Kai" was an Aubreita class sloop converted to appear as a merchant vessel for submarine trap (Q.ship) duty.

not be acquired in a short time. Command of "UB29" was taken over by Oblt. Platsch at the beginning of November and he brought a Dutch prize to Zeebrugge on November[174] 12th and 15th, it was lost on its first cruise to the Channel on December 6, 1916 it was destroyed by the English destroyer "Ariel" SW-Lich Bishop Rock with an explosive device.[175]

Hans Valentiner: - with "UB37"[176] only brought in the sinking of the brigantine "Twig" - 128 tons - 25 nm from Cape de la Hague; on November 6th he handed the boat over to Oblt. Günther. The latter sank two English fishing trawlers at Smith Knoll on his first Hoofden trip, during his first Channel cruise from November 26th to December 9th he sank 5 fishing trawlers at the Casquets (November 28th). He was shot at by a patrol boat while the latter were sinking. Günther had an artillery battle with a tanker on the 29th, but was still able to dive safely. When attacking a suspicious steamer there were heavy detonations close to the boat, apparently due to depth charges. He was shelled by French patrols on November 30th and by a large submarine on December 2nd. On December 1st and 4th at Ushant the Norwegian steamer "Erich Lindøe" – 1,087 tons - and the Greek steamer "Fofo" – 2,615 tons - were sunk, on December 6th 25 nm south of Eddystone Lightship, sank the Danish Steamer "Halfdan" – 1,307 tons - and finally 35 nm from Guernsey sank the French bark, “Marg. Dolfus "- 1,948 tons -. Total 7,147 tons. But it was Günther's last undertaking. On the next operation he was destroyed on January 14 by artillery fire and depth charges.[177]

After a few short trips, Waßner was on the way again to the Channel with "UB38" from November 12th to 16th. On November 13th, before Dieppe, he sank the English steamer "Caterham" – 1,777 tons -, 20 nm SSE from Beachy Head, also the English steamer "Bernicia" - 957 tons - and the French Barkentine "Riquette" - 165

---

[174] Typo in the original text. October indicated. Correct month indicated here.

[175] “UB29” was found in 2013 by divers off Flanders.

[176] Typo in original in which was printed “UB27”. Corrected in text.

[177] Typo in original which read February. Corrected above.

tons -. On 14 November south of the Isle of Wight the French steamer "La Roche Jaqueline"[178] – 2,200 tons -, north of Cape Antifer the Norwegian steamer "Melwang"[179] - 639 tons - and the French barkentine "Professor Jalaquier" 223 tons -, on November 13 north of Dieppe the French barkentine "St. Nicolas" - 300 tons - (together 6,222 tons). In addition, on November 14th, Waßner sank the French patrol vessel "Domince"[180] - 327 tons -; all in all an excellent result for a four day operation!

After Waßner Oblt. Amberger commanded "UB38" to a cruise to the Hoofden, he was in the Channel to Ushant from December 11 to December 22. His overall result was 10,614 tons, including the French Messageries Maritimes steamer "Magellan" - 6,027 tons - south of Beachy Head[181], the English fully rigged sailing ship "Najad" – 1,907 tons - on December 15, 25 nm SE-S from Bishop-Rock , the Spanish steamer "Ason" on December 17 – 2,083 tons - in front of the Bristol Channel. On another cruise from January 11 to January 24, 1917, which again returned to the Scillys and Ushant, Amberger achieved 4,798 tons, including the Spanish steamer "Manuel" – 2,419 tons - 85 nm west of Ushant (January 16) and the Norwegian steamer "Arp" – 1,759 tons - 45 nm from the Scillys. The English sailing vessel "Lillian H." - 467 tons at the “Lusitania’s” sinking site. On the outward journey, the boat got caught in a anti-submarine net from which it was just escaped as a patrol vessel approached. Strong patrols were also found on the return trip through Dover-Calais.

"UB39" sortied again under its proven commander, Oblt. Fürbringer, from the October 18th to the 30th to the western end of the Channel. Five steamers were sunk, including the English Steamer "Barbara" of 3,740 tons, 25 nm south of the Isle of Wight

---

[178] Credited to “UC17” according to p.351, vol. 3 of of the German Official History, Handleskrieg mit U-booten.

[179] Spelling is “Ullvang” in Norwegian consular documents.

[180] The claim was disallowed in vol3. p.295, Spindler op.cit.

[181] “Magellan” was sunk in December 1916 by a different submarine.

and the French steamers "Brigeny" (2,197 tons) and "Cannebière" (2,254 tons).[182]

Oblt. Küstner after two small cruises in the Hoofden (November 12th to 16th November, sinking result: collier "Corinth" – 3,669 tons; and November 20th to 22nd), proved to be a worthy successor to Fürbringer in two cruises in the western Channel.

On his first voyage into the western Channel from November 27 to December 11, 1916, Küstner sank 12 steamers and 3 sailing vessels, totaling 21,642 tons. He operated up to Start Point and from there around Ushant. Of particular note was the sinking of the French steamer "St. Philippe" west of Guernsey – 3,419 tons, which could only be overpowered by 40 artillery hits. The next operation from December 27, 1916 to January 11, 1917 led to the Bay of Biscay. The sinking of no less than 16 French fishing vessels was remarkable. A large transport steamer damaged in the dark near Cape de la Hague, on December 28, unfortunately escaped.

The list below shows which mine boats were active in Flanders during this war period, where they laid mines and what other sinking successes they achieved. - From this list it can be seen the small mine boats "UC1" to "UC11" apart from laying mines, could not carry out any trade warfare activity due to their low speed and limited radius of action and other applications as was possible with boats "UC16", "UC17", "UC18", "UC19", "UC21", "UC26" and "UC46". Among these results, those of Oblt. Kiel ("UC18") and Oblt. Saltzwedel ("UC21") with a sinking result of 31,721 tons and 36,234 tons respectively deserve special mention.

Correctly recording the vessels lost on the mines laid by Flanders boats is naturally difficult. In the accompanying compilation, only the results determined according to enemy reports (compilation of merchant ship losses) have been included.(Figs. 21-24 for minelaying areas and field placement)

---

182 "Barbara","Brigeny", and "Cannebière" were sunk in October 1916 by different submarines. "Brizeux" is the correct spelling for "Brigney". Fürbringer did sink five large steamers on this cruise.

## Flanders Mine Operations from October 8, 1916 to January 1917

| Boat | Cdr. Name | Sortie Date | Minefield location | Remarks |
|---|---|---|---|---|
| UC 1<br><br>12 mines per sortie | Oblt. Küstner | October 22 to 24 | Shipping route in front of Calais | |
| | | November 2 to 4 | SW of Dyck Lightship | |
| | Oblt. Thielmann | November 9 to 11 | West Dyck | |
| | | November 15 to 16 | Bergeness Bank-East Dyck | |
| | | November 22 to 24 | North of Calais | |
| | | November 30 to December 2 | " | |
| | | December 7 to 9 | NW of Gravelines | |
| | | December 14 to 16 | Haut Fonds de Gravelines. | |
| | Oblt. Steckelberg | December 27 to 28 | NW of Calais | |
| | | January 5 to 7 | At Calais | |
| | Oblt. Thielmann | January 13 to 15 | Dyck Lightship | |
| | | January 23 to 24 | At Calais | |
| | | January 31 to Feburay 2 | In front of Hewett Channel | February 1, dived when an enemy airship appeared |

Total number of mines laid: 156.

| Boat | Cdr. Name | Sortie Date | Minefield location | Remarks |
|---|---|---|---|---|
| UC 4 | Oblt. Buch | October 17 to 20 | Off Oxfordness | |
| 12 mines per sortie | Oblt. Howaldt | October 24 to 26 | Off Southwold | |
| | | November 4 to 9 | In front Lowestoft | November 9, attacked by enemy destroyers |
| | | November 23 to 25 | Off Oxfordness | |
| | Oblt. Reimarus | December 1 to 4 | North Galloper Buoy | |
| | | December 7 to 10 | Nort of Shipwash Lightship | |
| | | December 14 to 17 | In front of Stanford Channel | |
| | | December 27 to 29 | Off Oxfordness | |
| | | January 6 to 7 | At Sizewell Bank | |
| | | January 15 to 16 | Off Oxfordness | |
| | | January 22 to 24 | In front of Hewett Channel | |
| | | January 31 to February 2 | Off Oxfordness | February 2, dived when an enemy airship appeared. |

Total number of mines laid: 144.

| Boat | Cdr. Name | Sortie Date | Minefield location | Remarks |
|---|---|---|---|---|
| UC 6<br><br>12 mines per sortie | Oblt. Günther | October 13 to 15 | At Elbow Buoy | |
| | | October 21 to 23 | Before Queen and Endinburgh Channel | |
| | | November 3 to 4 | Elbow - Tongue | |
| | Oblt. Zerboni di Sposetti | November 10 to 11 | At Elbow | |
| | | November 15 to 21 | At N. Outer Gabbard Buoy. | Return voyage delayed due to severe engine malfunction |
| | | December 16 to 17 | At Kentish Knock. | |
| | | December 27 to 28 | At Shipwash Lightship | |
| | | January 5 to 7 | At Elbow Buoy | |
| | | January 14 to 15 | " | |
| | | January 19 to 20 | at Elbow | |
| | | January 23 to 24 | At Shipwash | January 23, sighted 3 English light cruisers and 3 destroyers. |

Total number of mines laid: 132.

| Boat | Cdr. Name | Sortie Date | Minefield location | Remarks |
|---|---|---|---|---|
| UC 11<br><br>12 mines per sortie | Oblt. Max Schmitz | October 17 to 20 | Shipwash. | |
| | | Octobe 24 to 27 | Kentish Knock. | |
| | | November 6 to 11 | In Black Deep. | |
| | | November 15 to 16 | At Sunk Lightship. | |
| | | November 21 to 23 | At Kentish Knock. | |
| | | November 27 to 30 | At Sunk Lightship. | November 28, English seaplane showdown by rifle fire, crew captured. |
| | Oblt. von Ditfurth | December 8 to 9 | Between Sunk and Shipwash Lightships. | |
| | | December 14 to 17 | At Sunk Lightship. | |
| | | December 27 to 28 | " | |
| | | January 6 to 7 | At Shipwash Lightship. | |
| | | January 14 to 15 | At Sunk Lightship. | |
| | | January 23 to 25 | At Shipwash. | |

Total number of mines laid: 144.

| Boat | Cdr. Name | Sortie Date | Minefield location | Remarks |
|---|---|---|---|---|
| UC 16 18 mines per sortie | Oblt. von Werner | October 17 to 20 | 3 mines in front of Folkstone. 3 mines at Royal Sovereign Lightship. 4 mines at Newhaven. 2 fields of 3 mines at Cape Beachy Head. | October 20, caught in a net. |
| | | November 7 to 19 | 6 mines at Royal Sovereign Lighthouse, 4 mines at Beachy Head, 8 mines at Nab Lightship. | November 8, enemy submarine sighted, no possibility of attack. Various artillery battles with steamers. The following were sunk: 1 steamer, 5 sailing vessels, 1 fishing trawler totaling 2184 tons |
| | | December 20 to 27 | 4 mines at Cape Dungeness. 2 mines at Royal Sovereign Lighthouse. 1 field at Beachy Head. 1 field at Owers Lightship. | December 25, began return journey due to diving plane damage. |
| | | January 16 to February 2 | In English Channel | 5 steamers and 1 sailing ship were sunk, a total of 7896 tons. |

Total number of mines laid: 144. - Total sinking result: 10080 tons (without mine successes)

May 1916 to February 1917

| Boat | Cdr. Name | Sortie Date | Minefield location | Remarks |
|---|---|---|---|---|
| UC 17 18 mines per sortie | Oblt. Wenninger | Octobe 21 to 26 | South of Flamborough Head. | October 25, started return trip due to stormy weather. |
| | | November 4 to 20 | 5 mines at Portland. 4 mines at Dartmouth. 5 mines at Plymouth. 4 mines at Falmouth. | November 8, pursued by an enemy destroyer. Artillery battles with various steamers(English And French). Sunk: 6 steamers, 5 sailing vessel, totaling 10070 tons. |
| | | December 14 to 30 | 3 mines at Shambles Lightship. 3 mines at Portland Bill. 4 mines at Berry Head. 4 mines at Eddyston-Plymouth. 2 mines at Falmouth. 2 mines at Manades Rock. | December 16 pursued by destroyers and trawlers. December 19 - shot at by a trawler. Sunk: 1 steamer, 8 sailing vessels, 2 schooners, total of 2613 tons. |
| | | January 18 to February 1 | 6 mines at Portland and 4 mines each at Dartmouth, Plymouth and Falmouth. | Sunk: 2 steamships, 3 sailing vessels, total 4858 tons |

Total number of mines laid: 72. - Total sinking result: 17541 tons (without mine successes)

| Boat | Cdr. Name | Sortie Date | Minefield location | Remarks |
|---|---|---|---|---|
| UC 18 18 mines per sortie | Oblt. Kiel | October 24 to 29 | Steamer route at Haisborough | |
| | | Novemer 4 to 19 | 2 fields at Boulogne, 1 at Brest. 3 mines at Les-Pierres-Noires. | November 9 and 13 - artillery battle with trawlers. Sunk: 1 French and 2 English steamers, a total of 5038 tons. |
| | | December 10 to 24 | At Brest, Nazaire and the mouth of the Loire. | Artillery battle with English steamer; attacked by the French destroyer, escaped by quick diving. During the operation, 8 steamers, 4 sailing vessels, a total of 12043 tons sunk. |
| | | January 11 to 25 | At Chenal-du-Four, Bay of Seine, Belle Isle. | Sunk: 7 steamers and 6 sailing vessels, a total of 14640 tons. |

Total number of mines laid: 72. - Total sinking result: 31721 tons (without mine successes)

| Boat | Cdr. Name | Sortie Date | Minefield location | Remarks |
|---|---|---|---|---|
| UC 19 18 mines per sortie | Oblt. Nitzsche | November 15 to 21 | 4 mines North of Newarp. 4 mines at Would Lightship. 5 mines to South of Haisborough. 5 mines at Foulness Buoy. | |
| | | 27-Nov | - | Sortied to place a field at Boulogne, but did not return. |

Total number of mines laid: 18.

This section intentionally left blank

| Boat | Cdr. Name | Sortie Date | Minefield location | Remarks |
|---|---|---|---|---|
| UC 21 18 mines per sortie | Oblt. Saltzwedel | November 26 to December 12 | 2 mines in front of Dungeness. 2 mines at Royal Sovereign Lightship. 2 mines at Owers Lighthship. 3 mines at the entrance to the Needles. | November 27 - was shelld by "submarine hunters". Artillery shootout with steamers. Sunk: 11 steamers, 6 sailing ships, 3 fishing schooners, total 22091 tons. |
| | | January 17 to 29 | At S. Nazaire at the mouth of the Gironde. | Sunk: 7 steamers, 7 sailboats, 3 fishing schooners, total 14143 tons. Forced to interrupt minelaying sortie due to a premature explosion in the gun barrel. |

Total number of mines laid: 32. - Total sinking result: 36234 tons (without mine successes)

May 1916 to February 1917

| Boat | Cdr. Name | Sortie Date | Minefield location | Remarks |
|---|---|---|---|---|
| UC 26 18 mines per sortie | Oblt. von Schmettow | September 28 to October 9 | Off Boulogne Lightship. On Le Harve - Cherbourg | Shot at by enemy steamers and patrols. Sunk: 3 steamers, 2 sailing vessels totaling 5518 tons |
| | | October 24 to November 1 | At Boulogne, Le Harve and Cherbourg | October 31, Operation canceled due to bad weather |
| | | November 13 to 27 | At the Bassure-de-Bas Buoy. At Le Havre and Cherbourg. | Shot at by patrols. Mining at Bordeaux cancelled due to bad weather, these mines were planted in front of Cherbourg on November 20th. The following were sunk: steamer, 3 sailing vessels totaling 15165 tons. |
| | | January 25 to 29 | At Inner-Dowsing. | - |

Total number of mines laid: 72. - Total sinking result: 20683 tons (without mine successes)

| Boat | Cdr. Name | Sortie Date | Minefield location | Remarks |
|---|---|---|---|---|
| UC 46 18 mines per sortie | Oblt. Moecke | December 5 to 9 | Steaner route East of Flamborough Head. Cape Spurn Point-Flamborough Line. | - |
| | | December 12 to January 3 | Dartmouth and Swansea | Battles with fishing trawlers and steamers. The following were sunk: 2 steamers, 1 sailing vesse; and 1 fishing trawler for a total of 4309 tons. The English steamer "William Middleton" (about 2000 tons)was torpedoed at Lundy-Island and towed to safety. |
| | | 25-Jan | - | Departed Zeebrugge for a mining operation in the Bristol Channel and did not return. |

Total number of mines laid: 36. - Total sinking result: 4309 tons (without mine successes)

May 1916 to February 1917

| Boat | Cdr. Name | Sortie Date | Minefield location | Remarks |
|---|---|---|---|---|
| UC 47 18 mines per sortie | Oblt. Hundius | January 25 to 30 | Off Flamborough Head | - |

Total number of mines laid: 18.

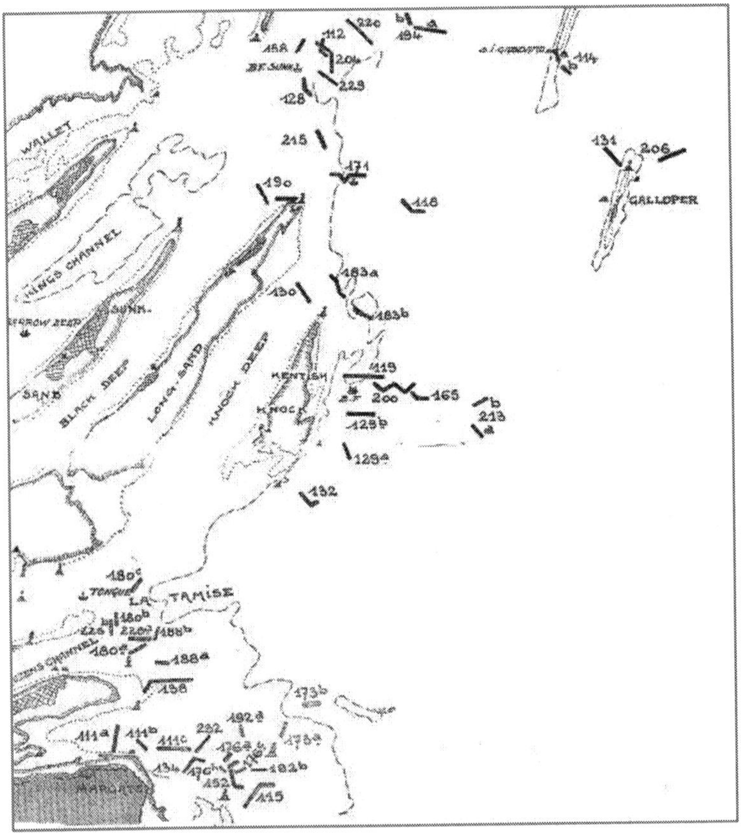


Fig 21. Examples of mine sorties to the mouth of the Thames. May 1916 to January 1917.

Fig 22. German mining of the English Channel coast.

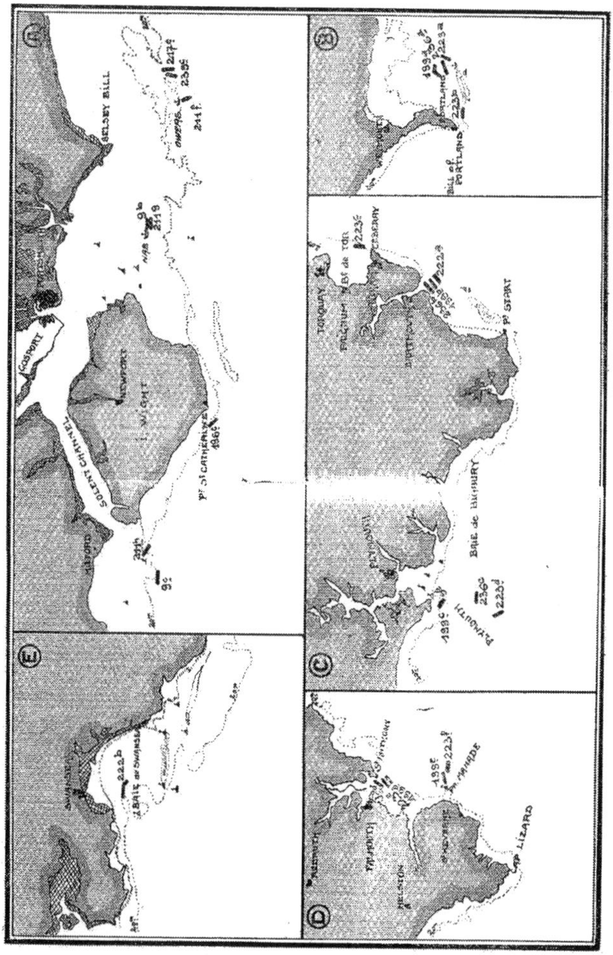

Fig 23. German mining off Normandy, Brittany and Gironde coast.

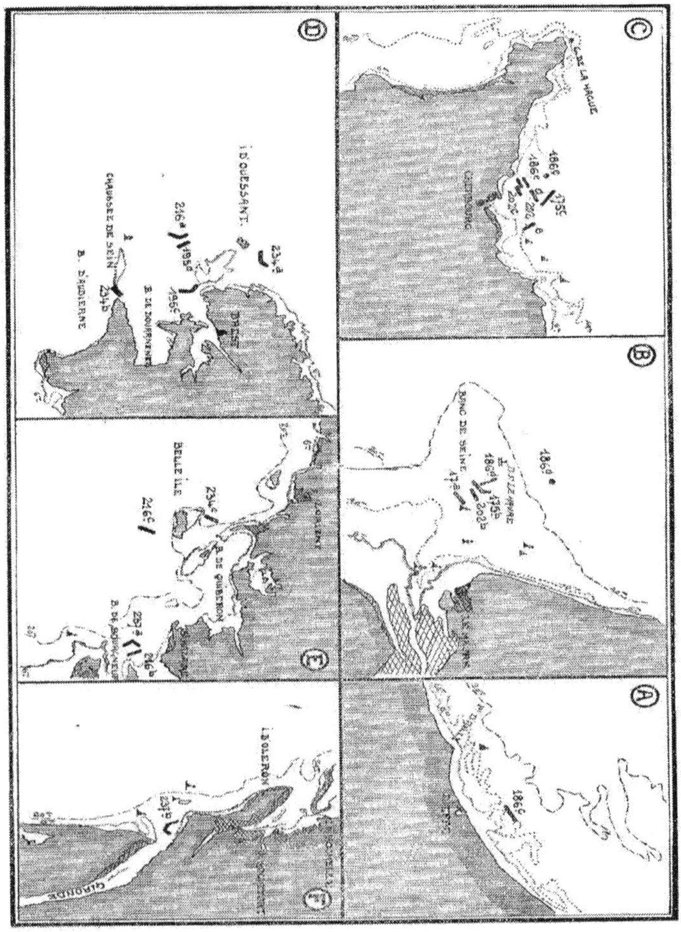

Fig 24. Mining of the coast between Boulougne and Cape Gris Nez and Dover and Newhaven, October 1915 to January 1917.

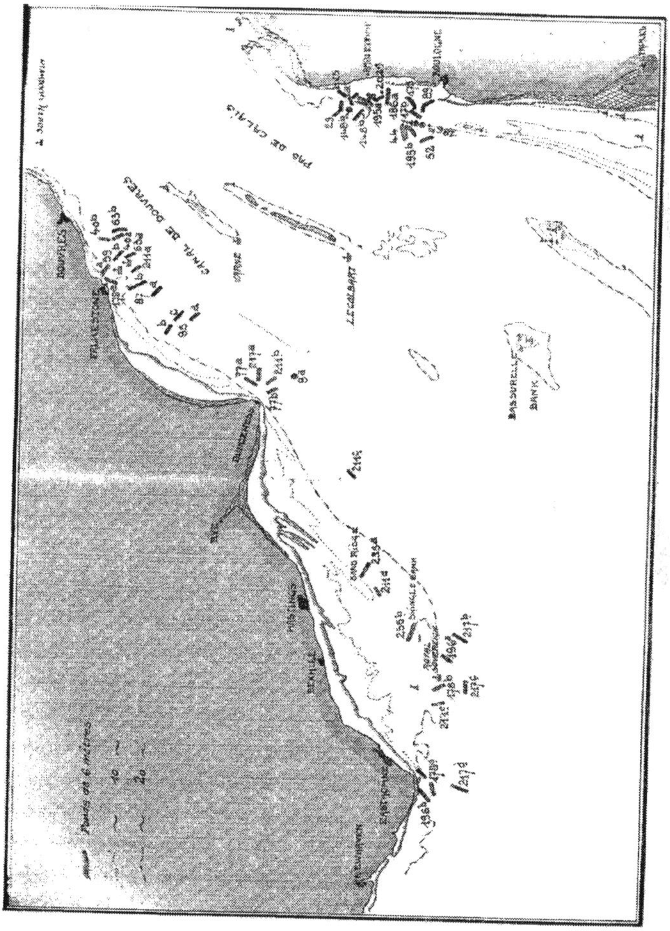

**List of merchant ships sunk from October 1, 1916 to January 1, 1917 on mines delivered by Flanders boats.**

| Name and class of ship | Flag | Gross Tons | Date | Mines placed by | Place of sinking |
|---|---|---|---|---|---|
| | | | 1916 | | |
| Fishing Trawler Farfadet | French | 17 | 7-Sep | Flanders boats* | - |
| Schooner Egberdine | Dutch | 167 | 8-Sep | " | - |
| Bark Theodor | Swedish | 643 | 19-Oct | " | North of the Channel |
| Steamer Fortuna | Dutch | 1254 | 22-Oct | " | - |
| Trawler Lord Roberts | English | 293 | 26-Oct | UC 11 | off Shipwash |
| Steamer Bygdo | Norway | 2345 | 27-Oct | UC 18 | In English Channel |
| Fishing trawler Blanc-Nez | French | 247 | 27-Oct | Flanders boats | - |
| Steamer Vanellus | English | 1797 | 1-Oct | UC 26 | Le Harve Roads |
| Steamer Mantola | " | ** | 30-Oct | UC 6 | At the mouth of the Thames |
| Fishing trawler Girl Eva | " | 76 | 2-Oct | UC 26 | - |
| Trawler Clenyrosen | " | 224 | 3-Nov | UC 6 | At Cross Sand |
| Trawler Cantatrice | " | 302 | 5-Nov | UC 4 | At Yarmouth |
| Steamer Lanterna | " | 1685 | 7-Oct | UC 1 | At Cromer |
| Fishing Trawler Eugenie | French | 38 | 8-Oct | Flanders boats | - |
| Mailsteamer Alaunia | English | 13405 | 19-Oct | UC 16 | Off Royal Sovereign Light |
| Steamer Huguenot | " | 1032 | 20-Oct | UC 11 | Off Sunk Lighship |

*"Flanders boats" is given in cases where it is not known exactly on whose mines the lost ships detonated. **Salvaged in the Thames (8,253 tons).

| Name and class of ship | Flag | Gross Tons | Date | Mines placed by | Place of sinking |
|---|---|---|---|---|---|
| From previous | | 23525 | 1916 | | |
| Steamer Framfield | English | 2510 | 24-Oct | UC 11 | Off Sunk Lighship |
| Steamer Sparta | " | 480 | 28-Oct | UC 4 | At Southwold |
| Fishing Trawler Lloyd George | " | 229 | 26-Oct | UC 11 | - |
| Hospital ship Galeka | " | 6772 | 28-Oct | UC 26 | At Cape de la Hague |
| Steamer Sunniside | " | 447 | 9-Nov | UC 4 | At Southwold |
| Minesweeper Fair Maid | " | 430 | 9-Nov | UC 4 | At Cross Sand |
| Minesweeper Ludlow | " | 810 | 29-Dec | UC 6 | At Shipwash |
| Lighter H. M. W. | " | 75 | 10-Nov | UC 18 | At Boulogne |
| Trawler Benton Castle | " | 283 | 10-Nov | UC 17 | At Dartmouth |
| Steamer Lake Michigan | " | * | 15-Nov | UC 18 | at Brest |
| Steamer St. Leonards | " | ** | 15-Nov | UC 26 | off Le Harve |
| Steamer Vasco | " | 1914 | 16-Nov | UC 16 | At Beachy Head |
| Trawler Anthony Hope | " | 288 | 16-Nov | UC 26 | At Le Havre |
| Fishing Trawler Burnley | " | 275 | 25-Nov | UC 4 | At Orfordness |
| Drifter Pelagia | " | 84 | 28-Nov | UC 16 | At Nab Lightship |
| Fishing Trawler Lord Airedale | " | 215 | 29-Nov | UC 11 | At Sunk Lightship |
| Fishing Trawler Camellia | " | 46 | 1-Dec | UC 17 | At Plymouth |
| Trawler Remarko | " | 245 | 3-Dec | UC 4 | At Lowestof |
| Steamer Tornate | Dutch | 5909 | 5-Nov | Flanders boats | - |

*Salvaged in Brest (9,288 tons). **Salvaged at Le Harve (4,574 tons).

May 1916 to February 1917

| Name and class of ship | Flag | Gross Tons | Date | Mines placed by | Place of sinking |
|---|---|---|---|---|---|
| From previous | | 44537 | 1916 | | |
| Steamer Poona | English | † | 6-Dec | UC 21 | - |
| Steamer Helene | Dutch | 1798 | 22-Nov | Flanders boats | - |
| Trawler Dhoon | English | 257 | 24-Nov | UC 19 | At Yarmouth |
| Steamer Themis | Dutch | 897 | ?-Dec | Flanders boats | - |
| Auxiliary Ship Tervani | English | 457 | 5-Dec | UC 4 | At Orfordness |
| Fishing Trawler Dagon | " | 250 | 8-Dec | UC 21 | At Royal Sovereign Lighthouse |
| Special Services Ship Kent County | English | 86 | 8-Dec | UC 4 | At Cross Sand |
| Steamer Strathalbyn | " | 4331 | 10-Dec | UC 26 | At Cherbourg |
| Steamer Harlington | " | 1089 | 9-Dec | UC 11 | Both 4 nautical miles SW of Shipwash |
| Steamer Harlyn | " | 1794 | 9-Dec | UC 11 | |
| Steamer Forth | " | 1159 | 9-Dec | UC 11 | - |
| Steamer Nora | Danish | 772 | 11-Dec | UC 46 | At Cape Spurn Point |
| Steamer Coath | English | 975 | 12-Dec | Flanders boats | - |
| Fishing Trawler Margaret | " | 54 | 17-Dec | UC 21 | At Cape Dungeness |
| Steamer Giustizia | Italian | 1169 | 3-Dec | Flanders boats | - |
| Steamer Modig | Norway | 1570 | 21-Dec | UC 46 | At Cape Flamborough Head |
| Trawler St. Ives | English | 325 | 21-Dec | UC 17 | Falmouth |
| Steamer Zoroastra | " | 3803 | 29-Dec | UC 11 | Off Sunk Lightship |
| Steamer Suffolk | " | * | 28-Dec | UC 16 | In front of Porsmouth |

† Salvaged in the Channel (7,626 tons) *Salvaged (7,573 tons).

| Name and class of ship | Flag | Gross Tons | Date | Mines placed by | Place of sinking |
|---|---|---|---|---|---|
| From previous | | 65323 | 1916 | | |
| Fishing Tawler St. Louis | French | 184 | 26-Dec | Flanders boats | - |
| Fishing Trawler Venus | " | 281 | 31-Dec | " | - |
| Steamer Lonada | English | 1286 | 29-Dec | UC 6 | North of Shipwash Lightship |
| Steamer Aspenleaf | " | * | 30-Dec | UC 16 | In English Channel |
| | | | 1917 | | |
| Trawler Donside | " | 182 | 7-Jan | UC 11 | At Lowestof |
| Steamer Sussex | " | ** | 1-Jan | UC 1 | Off Calais |
| Steamer Port Nicolsen | " | 8418 | 15-Jan | UC 1 | In front of Dunkirk |
| Trawler New Comet | English | 177 | 20-Jan | UC 4 | At Orfordness |
| Patrol ship ?? | French | 1000 | 20-Jan | UC 16 | - |
| Steamer Ole Bull | Norway | 1380 | 11-Jan | UC 18 | At Hasborough |
| Steamer Brabant | " | 1492 | 15-Jan | UC 46 | At Cape Flamborough Head |
| Fishing Trawler"Fleur de Genest" | French | 3 | 17-Jan | Flanders boats | - |
| Steamer Dagmar | Danish | 768 | 18-Jan | UC 4 | - |
| Steamer Kamma | Swedish | 1515 | 22-Jan | Flanders boats | - |
| Steamer Salland | Dutch | 3657 | 23-Jan | UC 17 | - |
| Steamer Argo | Norway | 1261 | 28-Jan | UC 26 | Both at Inner Dowsing Lighthouse |
| Steamer Heimand I | " | 506 | 28-Jan | UC 26 | |
| Steamer Epsilon | Dutch | 3211 | 2-Jan | UC 17 | - |

*Salvaged (7585 tons). **Salvaged (5686 tons).

May 1916 to February 1917

| Name and class of ship | Flag | Gross Tons | Date | Mines placed by | Place of sinking |
|---|---|---|---|---|---|
| From previous | | 90644 | 1917 | | |
| Trawler Holdene | English | 274 | 2-Feb | UC 4 | At Orfordnes s |
| Steamer St. Helene | French | 2128 | 1-Feb | Flanders boats | - |
| Steamer Modiva | Norway | 1276 | 31-Jan | UC 47 | At Cape Flamborou gh Head |
| Trawler Longset | English | 275 | 6-Feb | Flanders boats | At Cape Nells |
| Destroyer Ghurka | " | 880 | 8-Feb | UC 16 | Both at Royal Soveign Lightship |
| Steamer Lullington | " | 2816 | 8-Feb | UC 16 | |
| Steamer Foreland | " | 1960 | 12-Feb | UC 11 | 6 nautical miles south of Shipwash |
| Trawler Sisters Melville | " | 260 | 13-Feb | UC 4 | Off Aldeburgh |
| | | 100513 | | | |

When evaluating the success of these operations, this mine total of over 100,000 tons in four months can be safely increased if including the very large English ships which also hit Flanders mines but could still be salvaged. The most important of these ships are indicated in the above list; it turns out that about 50,000 more tons were lost to mine hits for several months after this period. It should also be noted in the preparation of the above tables listing ships striking mines and not sinking is only from English sources. What it meant to command and operate English minesweepers, drifters and trawlers at the time is clear enough from the compilation. These vessels were constantly on duty sweeping routes clear.

## Mediterranean and Black Sea

In the third volume we mentioned the director of submarine warfare in the Mediterranean, KKpt. Kophamel, had agreed to continue the trade war there according to Prize Rules. Accordingly, the activities of Mediterranean boats during the entire period from May 1916 to the end of January 1917 can be considered as a single unified campaign. The large extent of the successes can be seen from the compilation at the end of this chapter (pp.265-289), which is based on all available historical documents. It is not only related to the boats' reports themselves, but also all tangible compilations and publications of former enemy powers. It can be noted precisely from the latter that one has hardly been mistaken rgarding success against the enemy; the total result of around 804,000 tons speaks an eloquent language in this regard. The Imperial Naval Staff, who directed operations there rather closely, is particularly to be honored. The Naval Staff also quite rightly recognized, much sooner than submariners at the front, that we were ultimately waging a submarine war against global tonnage and it was necessary to strike with the best boats and commanders where we could expect the greatest sinking results with the minimum boat losses. It was therefore quite correct to dispatch "U32", "U63", "U64", "U65" and "U72" to the Mediterranean during this period when trade warfare stagnated in the North Sea. The boats "UB42", "UB43", "UB44", "UB45", "UB46" and "UB47", which were assembled in Pola, commenced operations in this period in the Mediterranean and Black Sea. "UC20", "UC22", "UC23", "UC34", "UC35", "UC37", "UC38" departed from home to the Mediterranean in this period up to the end of January 1917. It was unfortunate "U33" was also held in the Black Sea during this period and "U38" also had to operate from May to August 1916 there where, apart from bombarding some enemy coastal sites, only 4 steamers, 4 sailing vessels and 2 fishing trawlers were able to be sunk. Undoubtedly, in the Mediterranean Sea these boats would have achieved results at least ten times greater in the same period. Two operations Valentiner carried out after he was released from the Black Sea brought 49,987 tons (Western Mediterranean operating area) and 34,464 tons (Madeira operating area), but the main sinkings in the latter took place in the western Mediterranean. The French gunboat "Surprise" was also sunk at Funchal Roads on December 3 (see compilation p.279). These cruises felt akin to freedom. That Valentiner almost sank the Russian battleship "Imperatriza Maria" on June 44 with two torpedos is only a small consolation for the lack of success during his stay in the Black Sea. For tasks in the Black Sea one would have had to be content with B I and C I boats until B II and C II boats were available.

What the boats operational areas were during the trade war is evident from the exhaustive compilation mentioned above.

The following special operations, successes and events lying outside the trade war should be mentioned:

"U21" was attacked by an enemy submarine with torpedoes on May 7, 1916 while enroute from Cattaro to Pola. On October 22, it had an artillery battle with a submarine trap south of Greece.

"U32" - Hartwig - was taken under fire by a French battleship on 22 October 1916 in the area of Malta. The boat was considerably disturbed by enemy patrols after leaving Cattaro to operate on the Malta-Cerigo route 4 to 6 January 1917. On January 9, 1917, the English battleship "Cornwallis" (14,000 tons), escorted by destroyers, was sunk near Malta. Unfortunately, this operation had to be canceled due to steering gear damage and bad weather on January 16.

“U33" exited Constantinople from August 7 to 23, 1916, for a task in the Black Sea, to operate off Novorossisk, behind the enemy front at the Armenian coast and pick up at Anakria a special mission bound for the Caucasus. The boat was in Varna from the 19th to the 21st. On August 23, it entered the Bosphorus again. On September 7 Gansser bombarded Mangalia. While he was in Euxinograd from September 9 to 16, an enemy air raid on Varna took place. "U33" was able to shoot down an enemy aircraft. From September 16 to 22, operating off Constanta; sank two fishing cutters. On September 30th he went from Euxinograd back to the Black Sea for an operation off Sulina, south of Snake Island and on the Dobruja coast. There was a battle with a Russian torpedo boat on October 5; a damaged German aircraft was sunk after the crew was taken onboard. Skirmishes later took place again with a torpedo boat and three steamships. On October 9, "U33" returned to Varna.

Finally on November 15th, it departed Constantinople to Cattaro while conducting operations against enemy forces in the Aegean Sea. The technical condition of the boat made a basic repair absolutely necessary. The crossing was uneventful besides an artillery battle with an English steamer on November 21. The boat soon went from Cattaro to Pola's shipyard, where it had a major overhaul until April 11, 1917.

It should be noted repair facilities and capabilities in Pola and Cattaro were insufficient and could not cope with the requirements of such a large number of boats. A timely generous expansion, as occurred in Flanders (Bruges-Ostend) by KKpt. Bartenbach was unfortunately not done.

So it happened that in late 1916 and early 1917, when the operations in the Mediterranean were booming, boat repair time was increasing. Thus, in late 1916 and early 1917, when the trade war in the Mediterranean Sea was in full swing, boats wasted time standing idle as repair time extended more often. In 1918 this situation took on a threatening aspect. The Commander of the Submarine Office, Vice Admiral Ritter v. Mann, and von Tiechler, intervened here early in 1918 to resolve this situation, which will be described in the 6th volume.[183]

[183] Gayer did not publish the 6th volume. The 1st volume printed in 1920 envisioned five volumes. Apparently Gayer added a 6th volume to his planning when this volume appeared in 1930.

"U34" - KptLt. Rücker - got into a net barrage in the Otranto Straits the evening after leaving Cattaro on May 13, but was able to escape early on the 14th. The sinking of the steamers "Mira" and "Languedoc "(see list) only succeeded after fierce artillery battles.

On the next operation from August 17 to September 14, 1916, Rücker was tasked with picking up KptLt.Canaris from Spain in addition to attacking Italy to Saloniki transports in the Ionian Sea and conducting trade warfare in the western Mediterranean. On August 24, 25, 27, 28 and September 1, Rücker had artillery battles with steamers and patrols; one of the latter also tried to ram the boat. From August 29th to 30th the boat was at the rendezvous point in front of Cartagena, but no one was found.

Rückers third cruise in this period led to resumed trade warfare in the western Mediterranean. On October 30, the boat had an artillery battle east of San Sebastian with a submarine trap and was subsequently hunted by a submarine search group. On November 3 the pursuit of two steamers had to be abandoned due to return artillery fire from one of them.

We now describe Arnaud de la Perière's four cruises which achieved unprecedented results and were not superseded by Arnauld in the later unrestricted submarine war. These operations (see compilation) generated 57,148 tons, 90,213 tons, 66,747 tons and 16,610 tons[184], together around 230,000 tons. If we compare these cruises with Arnauld's four later cruises in 1917 during the initial phase of unrestricted submarine war, which yielded 25,170, 67,989, 29,509 and 25,106 tons, it becomes clear the odds of encountering countermeasures in 1916 compared to 1917 and in retrospect must state again the missed opportunities by suspending the trade war in the North Sea from April to October 1916. As other commanders,

[184] Respectively June-July, August-September, October-November, and January 1917. French forces targeted de la Perière. An example being the month of June was largely spent hunting him from June 7 to July 3, 1916; two squadrons of French torpedo boats, assisted by Italian vessels, crisscrossed the western Mediterranean basin, from the Balearics to Corsica and from Marseille to Bizerte, in an unsuccessful attempt to destroy him. He sank 35 vessels, while refueling in Cartagena, a neutral Spanish port, before returning to Cattaro.

Arnauld had countless artillery battles with patrols and steamers on his four trips. On his second voyage, he was even shot at from land (Cannes area) on August 12 while in combat with an Italian steamer. On August 17, the French armored cruiser "Ernest Renan" narrowly escaped Arnauld's torpedo in the Gulf of Lyon. On the third voyage, KptLt. Canaris and two other officers were picked up from Cartagena and the English small cruiser "Riegel" sunk (October 2, not far from the Balearic Islands).[185] A gun was captured from the English steamer "Bronwen" sunk on September 24, 25 nm NE from Dragonera Island, the cannon and important secret papers had been taken from the English steamer "Charterhouse" the day before. Of the 5 steamers sunk on the last operation from January 1 to January 13 (between Malta and Cerigo), 3 were overpowered after a long gun battle, 2 were torpedoed while submerged; 4 captains were taken prisoner. Due to heavy weather "U35" returned to Cattaro on January 13th.

The five operations of "U39", led by the old experienced campaigner Forstmann, resulted in about 170,000 tons and made the Mediterranean submarine cruiser war in 1916 one of the great successes we admire today. Forstmann's first cruise (May 16 to June 6 - 52,976 tons -) took place in the western Mediterranean. On May 23, the blast furnaces of Porto Ferrajo on Elba were bombarded with artillery; the fire was returned from land with rifles and artillery from Fort Stella. After the bombardment, "U39" was hunted by destroyers and depth charged. The sinking results on Forstmann's two succeeding cruises (July 5 to August 1 – 63,378 tons -, October 8 to November 1 – 17,296 tons -) are all the more astonishing since special tasks still had to be fulfilled on both trips. On the first trip mentioned, a German and Turkish mission had to be disembarked on the Tripoli coast while enroute to conduct trade warfare in the western Mediterranean and picked up on the way back. On July 8th and 9th anchoring was carried out in front of Jahoudie and the mission disembarked, on July 10th an Arab

[185] "Riegel" was an English built Arabis class sloop commissioned into the French Navy. It does appear as a small cruiser in profile.

sheikh was landed off Misurata. From July 13th to 24th the actual trade warfare portion was carried out. On July 13, before Bizerte, the boat was bombed by an airplane. On July 14th, 15th and 18th there were artillery battles with steamers and patrols. On July 20, battles off Algiers with a submarine trap, later with two destroyers and a patrol. Personnel from the missions were re-embarked on July 27 before Misurata and on July 28 before Jahoudie. On July 29th an artillery battle with the Italian troop transport steamer "Letimbro" (see list), which was subsequently sunk by torpedo. On August 8th, Forstmann returned to Cattaro.

In the following voyage (October 8th to November 1st) he anchored again on October 13th in front of Marsa Brega. Hauptman Sakib and ammunition were disembarked; the Todenwarth Expedition[186] and a man's corpse[187] were taken on board. On October 15, the entire cargo and all non-crew members were disembarked at Misurata. In the days that followed the gun was used often and it became unusable after a shell exploded in the gun barrel. As a result on October 22 Forstmann began the return voyage. On October 27 a Turkish major and the Todenwarth expedition were taken at Cape Misurata, on October 28 a Turkish captain and a sheikh who were to travel to Berlin were taken on board at Marsa Brega. On November 1, "U39" returned to Cattaro. The other two cruises led to the eastern Mediterranean. The sinkings (see compilation) mainly took place around 120 to 150 nm northwest of Alexandria.

On the second voyage from January 3 to January 27, 1917 - results were particularly limited because steamer routes had been considerably shifted; little traffic was encountered.

The passage of "U64" to the Mediterranean Sea has been already described on page 210. During this war period, the boat did not operate from Pola.

---

186 Intended to establish a wireless station at Misurata.

187 The remains of Otto-Felix Mannesmann.

"U65" - Fischel - had to abandon his first cruise from Cattaro after sinking the English steamer "Caledonia" 125 nm ES from Malta because the sinking steamer rammed and damaged the boat. After surfacing, a British major general[188], his aide and the steamer captain were captured and taken to Cattaro.

"U72" under KptLt. Krafft left Helgoland on August 20 at noon, planted mines before Lisbon on September 3, and off Oran on September 7 and at Bizerta on September 11. On the same day, the English steamer "Achaia" – 2,733 tons - ran into the mines 300 meters ENE of Oran's port entrance.

From November 16 to December 17, Krafft traveled to the Gulf of Genoa and produced 15,469 tons in sinking results which is significant for a large mine boat (see list). Krafft laid mines off Genoa on the afternoon of November 25 and off Marseille at noon on December 4. The Italian steamers "Audax" (2,130 tons) and "Lalento" (2,541 tons) and the English "Special Service ship Perugia" – 4,348 tons - (December 3) are believed to have fallen victim to this operation. [189]

Three mining operations Sieß carried out with the "U73" into the eastern Mediterranean are extremely significant. The first (July 24 to August 14, 1916) led to Salonika, to the Orphani Bay (east of Chalkidike) and the Bay of Mudros. It is not known whether these bays, which are widely used by the Entente, were regularly searched for mines, or whether large losses were avoided by happy coincidences on these very skillfully laid minefields. The next mine laying had tremendous consequences in the entrance to Salamis Bay (October 27), in the Kos passage (October 28) and in the Mykoni passage (October 29). The English hospital ship "Britannic" - 48,158 tons - ran onto a mine in the Kos passage on November 21, and the English hospital ship "Braemar Castle" – 6,318 tons - on

[188] Major General Hurdis Secundus Lalande Ravenshaw, G.O.C., 27th Division.

[189] Both "Audax" and "Lalento" went missing at sea while enroute to Genoa. Neither is credited to any submarine. "U63" was credited with sinking HMS "Perugia" in vol3. p.338, Spindler op.cit.

November 23th in the Mykoni passage. Apparently the mines were laid relatively deep, so only larger ships could strike them. While "Braemar Castle" could be salvaged, "Britannic" was completely lost. In addition, the mines in Salamis Bay probably sank a number of smaller Greek steamers.[190] The third venture led to off Alexandria and Port Said. Mines laid off Port Said on December 20 and Alexandria on the 22nd. The Greek steamers "Kiki Isaias" (2,993 tons), December 31, 1916, and "Argostolion" (594 tons), January 10, 1917, presumably sank on them.

After this voyage "U73" had to enter Pola's shipyard for repairs until September 14, 1917. After Krafft's journey to Marseille, "U72" also went to the shipyard until May 26, 1917. Both boats managed only one cruise throughout 1917, which once again confirms the inadequacy of repair facilities available at Pola and Cattaro for increased requirements.

It should be mentioned "U47" entered the repair yard upon arrival in Pola (December 31, 1916) and remained until May 1.

Among the successes of B-boats "UB42" to "UB47", those of "UB43" - Mellenthin - and "UB47" - Steinbauer - particularly stand out well in this period. In four trips (See list) a grand total of 63,243 tons!

"UB42" - Wernicke -spent itself from June 8 to 30, 1916 in a vain attempt to travel from Pola to the Dardanelles. On June 18, the boat was at "Hersing's Stand" (Gallipoli Peninsula outside the Dardanelles) when news was received indicating entry was impossible due to the danger of mines. After Wernicke was bombarded by a battery at the "Hersing's Stand" on the 19th, since his boat was thought to be English, he set off back to Cattaro.

On the next trip, entry into the Dardanelles was successful on August 15th. Wernicke was in Constantinople on August 16. He later operated in the Black Sea on five different cruises in a manner similar to "U38", where he was based at the Bulgarian port of

[190] Footnote in original. Steamer "Angeliki" October 29, 1916 (677 tons), steamer "Sparti" November 21, 1916 (961 tons).

Varna. Apart from the sinkings mentioned (in the table on p. 285) and towing a friendly aircraft on September 27, no special events were recorded. On September 11th and 12th enemy light forces were sighted in front of Constanza, on October 7th two transports accompanied by torpedo boats in front of Penekop Bay; Wernicke did not attack in either case. On October 19, a torpedo hit was scored on a transport between Cape Midia and Sevastopol. On a further attack attempt the boat ran under the steamer and bent its periscope on the bilge keel. From the end of 1916 to February 3, 1917, the boat could not be used on operations.

"UB43" achieved its aforementioned excellent results despite the usual artillery battles with armed steamers. His area of operations was always the Malta – Cerigo passage, and on the last voyage from November 9 to 17, the Cretan Sea. On October 8, 1916, Mellenthin scored a torpedo hit on an Italian small cruiser, which the latter survived.[191]

"UB44" under Oblt. Wager, one of our oldest submarine officers was dogged by misfortune. When returning from its first cruise on July 9, the boat was shot at by an enemy submarine with torpedo when entering Cattaro; on the next venture, which was to lead to the Dardanelles, the boat was caught by the English guard drifters "Quarry", "Knowe" and "Garrigill" in the Otranto Street exit and destroyed with depth charges.[192] It was the only Mediterranean submarine loss in this war period.

The next boat, "UB45", under KptLt. Palis, was lost in the Black Sea. After doing one operation from Cattaro in the Mediterranean and one from Constantinople to the Black Sea - on the latter scoring a torpedo hit a Russian transport steamer (before Trebizond August 29) - it ran into a mine on November 6th 2 1/2 hours after leaving Varna and sank.

---

[191] This claim was disallowed in Spindler, op. cited. P.341.

[192] The loss by these means has been disputed by newer research.

Also "UB46", KptLt. Bauer, who had operated in the Mediterranean from July 19 to August 12, successfully reached Constantinople on the next trip on October 7, then operating from Varna between Sevastopol and the mouth of the Danube only sank a Russian sailing vessel of 116 tons (!) after an attack approach failed on a Russian small cruiser escorted by two destroyers on October 31, he then ran into a mine on the next trip as it entered the Bosphorus and sank.

Thus the Black Sea considerably eroded our submarine war without achieving any significant military successes.

"UB47", Oblt. Steinbauer, which was fortunately left in the Mediterranean and became one of the "big ones" with a sinking result of 57,237 tons, especially if we take into account its success in other areas.

On October 16, Steinbauer, per orders. fired 60 shells at the factory and railway facilities at Marina di Catanzaro in the Gulf of Squillace (Italy). The most beautiful "end of the year", however, was the sinking of the French battleship "Gaulois" – 11,300 tons - on December 27 near Cape Maleas (Eastern Cape of the Peloponnese). The sinking was achieved despite being escorted by small cruisers and fishing trawlers. After the attack, Steinbauer was violently attacked by the escort vessels with depth charges and hunted by motor boats for six hours.

We must mention the activity in this period of the brave small boats UB-14 and UC-15 in the Black Sea. "UB14" was under Heimburg, well known for his earlier successes, from May 25 to June 8 off to the Laristan coast to attack Russian warships and transports. During the venture, Kavak, Zunguldack, Kara Agatsch Bay, Sinope and Kera Sund were visited. Under the command of Oblt. Schwarz, the boat made six more trips in the Black Sea. On July 4, a Russian battleship could not be sunk off Sevastopol due to a strong escort. On August 5, two Russian battleships were sighted in front of Sevastopol, again with no possibility of attack. On November 16, the last operation - defending Constanza against bombardment by the

Russians - had to be canceled due to a motor accident. "UB14" was then in Constantinople under repairs until May 28, 1917.

"UC15" had planted three mine fields in the fairway from Odessa to Nikolayev under the tried and tested command of Oblt. v. Dewitz from May 25 to 31, 1916. From June 5 to November 5, the boat was docked in Constantinople. On November 6, under Oblt. Heller it departed to transport mines to Varna; receiving orders on November 7th to turn back before Varna due to mine danger from new fields detected there.

With this boat the Black Sea claimed its third victim in this war period. He did not return from a mining operation Heller carried out into the northwestern Black Sea on November 13.

"UC20" (the characteristics of these newest boats of this war period are described on page 199) under Obl. Franz Becker departed from Helgoland to the Mediterranean on October 18. According to Naval Staff orders, five officers and two men from the General Staff and war material were to be transported to western Morocco. On November 14, two officers, two men and some cargo were disembarked with great difficulty off Wadi Draa (west coast of Morocco).[193] On November 15th, he steered towards the sea to Las Palmas to capture a vessel to serve as a lighter. However, this was not possible; the Portuguese sailing vessel "Emilia" (1,159 tons) was sunk on November 17 in front of Las Palmas. Returning to the African coast, 80 rifles and about 10,000 cartridges were rowed ashore in a dinghy on November 23.

Further attempts at landing and unloading material were abandoned due to lack of oil. The adventurous journey came to an end with arrival in Cattaro on December 11th. Danish sailing vessel "Fritz Emil"[194] (194 tons) was sunk in the North Sea On October 19. "UC20" had no mine shafts; the space gained was used to

---

[193] Hauptmann der Reserve Dr. Pröbster of the NfO intelligence bureau led the mission to foment an uprising in Morroco.

[194] Typo in original which printed "Emila".

accommodate cargo. In 1917, six further transport trips of a similar type were carried out.

"UC22" under the old Mediterranean and Black Sea commander von Heimburg went from Helgoland around the Shetlands to the Mediterranean on September 22nd. On September 28, the Russian sailing vessel "Emma" of 279 tons was sunk off the west coast of Ireland; on October 12th Heimburg arrived in Cattaro.

Apart from the sinking result of 22,902 tons mentioned in the list, Heimburg planted mines in front of Marsa Scirocco (Malta) on November 29 and the same on December 27. On December 29, a minefield was also laid in front of St. Paul (Malta). On November 29, the steamer "Luciston" – 2,948 tons - chartered by the English government ran into the mines in front of Marsa Scirocco (included in the above result).

From January 21 to January 25 Heimburg was minelaying at Brindisi in collaboration with the Austrian submarine "U62". On January 23, two minefields were planted there guided by observing shipping traffic in the wake of a minesweeper.

"UC23" - Oblt. Kirchner - departed from Helgoland on September 13th to Cattaro and arrived there on October 13th. Kirchner's crossing to Constantinople was a great success as the mines he laid at the entrance to Suda Bay (western part of the island of Crete) on November 24th hit the English steamer "Minnewaska" displacing 14,317 tons on November 29th. After mines had been laid in front of Candia on November 25th and on November 26th in front of Milos (southwestern island of the Cyclades), Kirchner continued to the Dardanelles and arrived in Constantinople on December 6th.

"UC34" - Sprenger -, which arrived in Cattaro on January 8, 1917, from home, did not operate in this war period; likewise "UC35" - Ernst v. Voigt - who arrived on December 25th.

"UC37" - Oblt. Launburg – departed Helgoland on December 23, 1916, driving through Dover Straits; he arrived in Cattaro on

January 11, 1917. He achieved the excellent result of 21,080 tons during the passage as detailed in the compilation.

The small boat "UC14", which we last saw (page 157) under the command of Obl. Franz Becker (later commander of "UC20") mined the steamer routes near Corfu, undertaking this activity under Oblt. Klatt from the beginning of August to mid-October 1916.

Klatt laid 12 mines each:
On August 7th South of Maestra Light Ton,(Po River mouth)
- 20th – South of San Vito Lighthouse in Taranto roads,
- September 6th west Cape Sta. Maria di Leuca,
- 21st South of Cape Sta. Maria di Leuca,
- October 6th off Gallipoli in the Gulf of Taranto.

Subsequently "UC14" was decommissioned on November 14, 1916 in Pola. It was dismantled, transported to Ostend and put back into service there in January 1917.

These mines are certain to have sunkt: the English admiralty drifter "Michaelmas Daisy" (99 tons) on November 26, 1916 near Cape Sta. Maria di Leuca, as well as the drifter "Finroß" (78 tons) on the same day before Gallipoli (Gulf of Taranto). The sinking of the English collier "Inververbie" – 4,309 tons - in the Gulf of Squillace on September 14 is probably also due to the "UC14's" mines.[195]

We have now reached the end of the description of the Mediterranean warfare in this war period, which represents the most successful period of any in the submarine war. The results described above per boat per operation were not obtained again in the unrestricted submarine war phase. The Commander of Mediterranean Submarines in this period, KKpt. Kophamel, had quite correctly recognized in the Mediterranean in this period it was not a matter of whether submarine cruiser warfare or unrestricted submarine warfare was waged, but only concerned that one struck

---

[195] Revised assessements credit the Austro-Hungarian submarine "U4" with the sinking.

in every possible manner while the enemy's countermeasures were still limited.[196] What was true of the Mediterranean Sea would also have been true of all boats that were able to work more freely in areas outside of more extensive patrols: in the Atlantic Ocean, on the Spanish and Portuguese coasts and west of Ireland.

The only sinking in the Mediterranean Sea that led to a relatively harmless exchange of notes with America during this era was the torpedoing without warning of the transport "Arabia" on November 6 by Mellenthin near Cape Matapan.[197] However, this did not have any significant influence on the activity of the Mediterranean submarines.

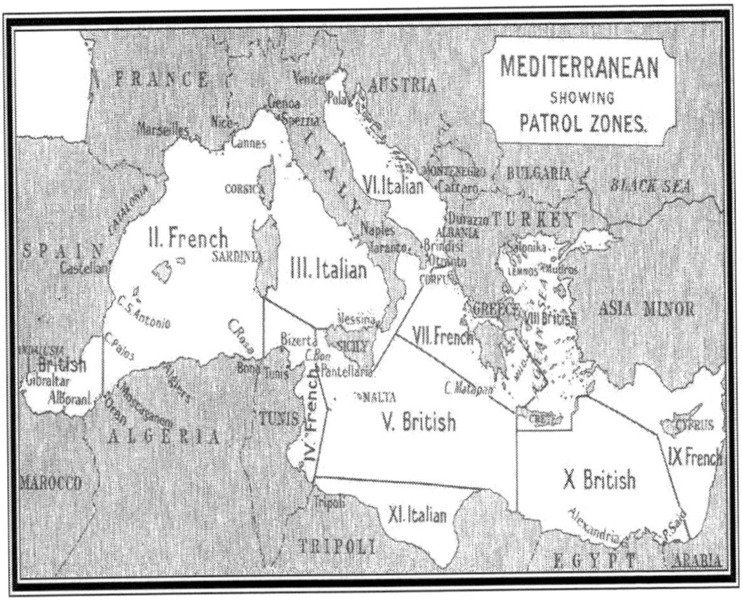


Fig 25. Patrol Zones

---

[196] Demonstrating the fractured Entente response; In Spring 1916 (Britain, France, Italy) agreed to divide the Mediterranean into 11 patrol zones. Fig 25. image sourced from Henry Newbolt, Naval Operations, Volume 4, June 1916 to April 1917, London: Longmans, Green, 1928.

[197] Germany claimed the ship was darkened and traveling on the Malta-Cerigo military transport route.

**Results of submarine operations against against trade in the Mediterranean and Black Sea May 1916 to January 1917. (Excluding mine losses)**

| Vessel class and name | Flag | Month and day | Location where Sunk | Tons |
|---|---|---|---|---|
| **U 21 - KptLt. Hersing - October 16 to November 12, 1916.** | | | | |
| Sailing Vessel Marina G. | Italian | 26-Oct | 27 miles from Pantellaria | 154 |
| Sailing Vessel Gilda R. | " | 28-Oct | 25 m. SW from Cape Chiappa | 34 |
| Brig Tre Fratelli D | " | 28-Oct | 30 m. SW from Cape Chiappa | 190 |
| Transport Glen Logan | English | 31-Oct | 10 m. SE from Stromboli | 5838 |
| Steamer Torero | Italian | 1-Nov | Near Ustica | 767 |
| Steamer Renardo Canal | " | 1-Nov | North coast of sicily | 1346 |
| Sailing Vessel S. Antonio | " | 3-Nov | At Cape St. Vito | 114 |
| Sailing Vessel San Giorgio | " | 4-Nov | 90 m. SE from Cape Carbonare | 258 |
| Transport No. 7 | " | 5-Nov | South of Malta | 12000 |
| | | | | 20701 |
| **U 21 - KptLt. Hersing - December 9 to 31, 1916.** | | | | |
| Transport Thistleban | English | 23-Dec | 5 m. NNW from Alexandria | 4117 |
| **U 32 - Hartwig - November 23 to December 14, 1916.** | | | | |
| Transport City of Birmingham | English | 27-Nov | 90 miles from Malta | 7498 |
| Transport Karnak | French | 27-Nov | Malta - Cerigo | 6816 |
| Two Masted Bark Celeste | Italian | 30-Nov | Marittimo Island | 1800 |
| | | | Continued next | 16114 |

| Vessel class and name | Flag | Month and day | Location where Sunk | Tons |
|---|---|---|---|---|
| | | | Previous page | 16114 |
| Sailing Vessel Lampo | " | 1-Dec | At Janutri Island | 59 |
| Sailing Vessel Cuore di Gesu | " | 1-Dec | At Janutri Island | 199 |
| Sailing Vessel Angela Madre | " | 2-Dec | At Elba Island | 155 |
| Tanker Lucellum | English | 3-Dec | Made Villefranche despite torpedo damage | - |
| Steamer Campania | Italian | 4-Dec | Bastia | 4297 |
| Sailing Vessel Carmentina Dominici | " | 8-Dec | Paola | 94 |
| Steamer Ursula | English | 12-Dec | 45 miles from Malta | 5011 |
| | | | | 25929 |

**U 32 - Hartwig - January 4 to 18, 1917.**

| | | | | |
|---|---|---|---|---|
| Steamer Rosalia | Italian | 7-Jan | 35° 11' N , 16° 48' East | 7186 |

**U 34 - Rücker - May 12 to June 6, 1916.**

| | | | | |
|---|---|---|---|---|
| Steamer Mira | French | 15-May | Malta Channel | 3050 |
| Brigantine San Andrea | Italian | 18-May | At Cape Feno | 224 |
| Steamer Adamantios Corais | Greek | 19-May | - | 2947 |
| Steamer Languagesedoc | French | 20-May | 45 miles SW of Marseille | 1612 |
| Brigantine Fabbricotti | Italian | 20-May | 40 miles from Cape San Sebastian | 150 |
| | | | Continued next | 7983 |

| Vessel class and name | Flag | Month and day | Location where Sunk | Tons |
|---|---|---|---|---|
| | | | Previous page | 7983 |
| Bark Erminia | " | 20-May | 45 miles SW of Marseille | 1544 |
| Steamer Tjomo | Norway | 21-May | In the open sea | 1454 |
| Barkentine Myosotis | French | 21-May | " | 356 |
| Bark Roberto G | Italian | 22-May | 45 miles SSE of Barcelona | 587 |
| Bark Regina | Russian | 22-May | 45 miles SSE of Barcelona | 593 |
| Bark Genista | Italian | 22-May | 45 miles from Barcelona | 1856 |
| Steamer Istros | Greek | 22-May | 45 miles from Barcelona | 1891 |
| Square Rigger Orealla | Italian | 22-May | Presco Tarragona | 1876 |
| Square Rigger Australia | " | 22-May | 60 miles SE of Balearic Islands | 1585 |
| Steamer Cornigliano | " | 23-May | At the Columbretes Islands | 2862 |
| Steamer Julia Park | English | 30-May | 40 miles from Cape Bougaroni | 2900 |
| | | | | 25487 |

**U 34 - Rücker - August 17 to September 14, 1916.**

| | | | | |
|---|---|---|---|---|
| Brigantine Maria | Italian | 21-Aug | At Cape Spartivento | 242 |
| Schooner S. Pietro | " | 22-Aug | South of Marittimo | 53 |
| Schooner Alisa | " | 24-Aug | At La Galite | 141 |
| Schoone Angelina | " | 24-Aug | Lampedusa | 43 |
| Steamer Cocoa | French | 25-Aug | In front Bougie Gulf | 2772 |
| Square Rigger Tarridon | Italian | 27-Aug | 25 miles SE of Formentora Island | 1526 |
| | | | Continued next | 4777 |

May 1916 to February 1917

| Vessel class and name | Flag | Month and day | Location where Sunk | Tons |
|---|---|---|---|---|
| | | | Previous page | 4777 |
| Bark Gorgona | " | 28-Aug | 20 miles NE of Cape Palos | 861 |
| Bark Tede | " | 29-Aug | East of Cartagena | 1273 |
| Bark Nostra Signora Signora Assunta | " | 31-Aug | 30 miles E of Cape St. Antonio | 256 |
| Bark Santa Maria | " | 31-Aug | Ibiza | 947 |
| Bark Quinto | " | 31-Aug | 35 miles E of Cape St. Antonio | 836 |
| Brigantine Guiseppe | " | 1-Sep | 30 miles from Ibiza | 180 |
| Steamer Baron Yarborough | English | 1-Sep | East of the Columbretes Islands | 1784 |
| Barque Pasquale Lauro | Italian | 4-Sep | South of the Gulf of Lyon | 1188 |
| Bark Selverstream | " | 4-Sep | In the open sea | 1124 |
| Bark Luigia | " | 7-Sep | 42 miles NW of Ajaccio | 917 |
| Steamer Elisabeth IV | Norway | 8-Sep | In the open sea | 7395 |
| Bark Spiridon | Greek | 10-Sep | South of Cape Spartivento | 562 |
| Bark Ellin | " | 10-Sep | " | 631 |
| Bark Panagia Akathistou | " | 12-Sep | East of Malta Channel | 421 |
| | | | | 23152 |

| Vessel class and name | Flag | Month and day | Location where Sunk | Tons |
|---|---|---|---|---|
| **U 34 - Rücker - October 19 to November 13, 1916.** | | | | |
| Schooner Valborg | Danish | 26-Oct | 7 miles South of Bougie | 207 |
| Steamer Germaine | Greek | 28-Oct | Denia | 2573 |
| Schooner Maria Theresa | French | 29-Oct | Columbretes Islands | 219 |
| Schooner Giavanni Anteri Beretta | Italian | 2-Nov | At Cape Tortosa | 332 |
| Steame Mogador | French | 4-Nov | Between Cape Tortosa-Mallorca | 1364 |
| Steamer Sheldrake | English | 8-Nov | At Adventure Bank | 2697 |
| Steamer Mudros | Italian | 8-Nov | " | 3137 |
| | | | | 10529 |
| **U 35 - von Arnauld - June 1 to July 3, 1916.** | | | | |
| Steamboat Motia | Italian | 13-Jun | Near Ustica | 500 |
| Sailing Vessel Maria D. Torre del Greco | " | 13-Jun | 120 miles East of Ustica | 77 |
| Sailing Vessel ? | " | 13-Jun | Near Ustica | 200 |
| Sailing Vessel Giome-Torre del Greco | " | 14-Jun | West of Poza | 20 |
| Sailing Vessel Francesco Gaeta | " | 14-Jun | 50 miles SSW of Capri | 43 |
| Sailing Vessel Antonio Chiogia | " | 14-Jun | Palo | ? |
| Steamer Tavolara | " | 14-Jun | 20 miles Eat of Civita | 701 |
| Sailing vessel Adelia | " | 15-Jun | At Elba | 170 |
| Sailing Vessel Anetta | " | 15-Jun | In the Ligurian Sea | 112 |
| Bark G.Maria | " | 15-Jun | " | 515 |
| Steamer Sardinia | English | 15-Jun | 38 miles from Gorgona Island | 1119 |
| | | | Continued next | 3457 |

May 1916 to February 1917

| Vessel class and name | Flag | Month and day | Location where Sunk | Tons |
|---|---|---|---|---|
| | | | previous page | 3457 |
| Sailing Vessel Audace | Italian | 15-Jun | 3 miles NW of Cape Corse | 143 |
| Sailing Vessel Dolmetta | " | 16-Jun | At S. Remo | 48 |
| Sailing Vessel Rondine | " | 16-Jun | At Port Maurizio | 112 |
| Bark Eva | " | 16-Jun | 6 miles from Port Maurizio | 1077 |
| Steamer Gaffa | English | 16-Jun | 80 miles SSE of Genoa | 3922 |
| Sailing Vesssel Eufrasia | Italian | 16-Jun | 10 miles North of Capraia Island | 71 |
| Steamer Poviga | " | 17-Jun | 72 meters from the Gigra Islands | 3360 |
| Steamer Rona | English | 17-Jun | 90 miles West of Cape Falcon | 1312 |
| Steamer Beachy | " | 18-Jun | 98 miles from Port Mahon | 4718 |
| Stamer Olga | French | 18-Jun | North of Minorca | 2964 |
| Steamer Aquilla | Norway | 18-Jun | " | 2190 |
| Sailing Vessel Maria C. | Italian | 19-Jun | 30 miles North of Port Mahon | 398 |
| Sailing Vessel France et Russie | French | 19-Jun | West of Minorca | 329 |
| Steamer Herault | " | 23-Jun | Gulf of Valencia | 2298 |
| Sailing Vessel Guiseppina Trapani | Italian | 23-Jun | SE of Vinaros | 1861 |
| Bark Saturnina | " | 24-Jun | 16 miles SE of Barcelona | 1568 |
| | | | Continued next | 29828 |

| Vessel class and name | Flag | Month and day | Location where Sunk | Tons |
|---|---|---|---|---|
| | | | previous page | 29828 |
| Sailing Vessel San Francesco | " | 624 | 25 miles S of Barcelona | 1059 |
| Sailing Vessel Chechina | French | 24-Jun | East of Barcelona | 177 |
| Steamer Dayeton Maru | Japan | 24-Jun | " | 3184 |
| Steamer Canford Chine | English | 24-Jun | 5 miles from La Calella | 2398 |
| Steamer Fournel | French | 25-Jun | North of Minorca | 2047 |
| Steamer Clara | Italian | 25-Jun | At Marittimo | 5503 |
| Steamer Pino | " | 27-Jun | 6 miles from Port Mahon | 1677 |
| Steamer Roma | " | 27-Jun | 50 miles from Minorca | 2491 |
| Steamer Mongibello | " | 27-Jun | 50 miles from Port Mahon | 4059 |
| Steamer Windermere | English | 27-Jun | 58 miles from Port Mahon | 2292 |
| Sailing Vessel Guseppina Via | Italian | 29-Jun | 50 miles SE of Cape Carbonara | 214 |
| Steamer Teano | English | 29-Jun | 24 miles NW from Marittimo | 1907 |
| Sailing Vessel Carlo Alberto | Italian | 29-Jun | 10 miles NW from Marittimo | 312 |
| | | | | 57148 |

**U 35 - von Arnauld - July 26 to Augsut 20, 1916.**

| | | | | |
|---|---|---|---|---|
| Steamer Dandalo | Italian | 28-Jul | At Cape Spartivento | 4977 |
| Sailing Vessel Sab Guiseppe Marta | Tunisian | 30-Jul | Sicilian Straits | 111 |
| Steamer Ethelbryhta | English | 30-Jul | 11 miles WSW from Pantellaria | 3084 |
| | | | Continued next | 8172 |

May 1916 to February 1917

| Vessel class and name | Flag | Month and day | Location where Sunk | Tons |
|---|---|---|---|---|
| | | | previous page | 8172 |
| Steamer Katholm | Danish | 30-Jul | Malta | 1324 |
| Steamer Britannic | English | 30-Jul | 20 miles from Cape Bon | 3487 |
| Sailing Vessel Generale Ameglio | Italian | 31-Jul | 60 miles from Kani Island | 222 |
| Steamer Citta di Messina | " | 31-Jul | Off Marittimo | 2464 |
| Steamer Erling | Norway | 31-Jul | In the open sea | 122 |
| Steamer Einar | " | 31-Jul | " | 136 |
| Sailing Vessel Emilio G. | Italian | 31-Jul | 60 miles from Kani Island | 166 |
| Steamer Heighington | English | 1-Aug | 40 miles NE from Cape Serrat | 2800 |
| Sailing Vessel Eugenia | Italian | 2-Aug | Southern tip of Sardinia | 550 |
| Sailing Vessel Neptune | French | 2-Aug | " | 151 |
| Steamer Tricoupis | Greek | 3-Aug | Gulf of Lyon | 2387 |
| Steamer Favonian | English | 4-Aug | 24 miles SW from Planier Island | 3049 |
| Steamer Lottenham | " | 4-Aug | 33 miles SWW of Marseille | 3106 |
| Steamer Siena | Italian | 4-Aug | 43° N, 5° E | 4372 |
| Steamer Teti | " | 4-Aug | 45 miles SE from Planier Island | 2868 |
| Steamer Mount Coniston | English | 5-Aug | 7 miles ES of Meda Island | 3018 |
| Steamer Achilleus | Greek | 5-Aug | At Cape San Sebastian | 843 |
| Steamer Trident | English | 7-Aug | 34 miles NE 1/4E from Dragonera | 3129 |
| | | | Continued next | 42366 |

| Vessel class and name | Flag | Month and day | Location where Sunk | Tons |
|---|---|---|---|---|
| | | | previous page | 42366 |
| Steamer Newburn | " | 7-Aug | " | 3554 |
| Sailing Vessel Speme | Italian | 7-Aug | 41° 57', 4° 31' E | 1229 |
| Steamer Imperial | English | 8-Aug | 38 m on SW-S from Planier | 3818 |
| Steamer Ganekogorta Mendi | Spanish | 9-Aug | Gulf of Lyon | 3061 |
| Steamer Sebastiano | Italian | 9-Aug | 40 m. From Cape San Sebastian | 3995 |
| Steamer Antiope | English | 9-Aug | 88 miles SWW from Marseille | 2973 |
| Steamer Tenriu Maru | Japan | 10-Aug | Gulf of Lyon | 3224 |
| Steamer Pagassari | Spanish | 11-Aug | " | 3287 |
| Sailboat Gina | Italian | 12-Aug | 35 miles East of Îles d'Hyères | 443 |
| Steamer Nereus | " | 12-Aug | 4 miles off Cape Garoupe | 3980 |
| Sailing Vessel St. Gaetan | French | 12-Aug | South of Cannes | 125 |
| Sailing VesselRegina pacis | Italian | 12-Aug | 7 miles South of Villefranche | 2227 |
| Sailing Vessel Eurasia | " | 13-Aug | 15 miles SSE of Cape Mele | 1898 |
| Sailing Vessel Balmoral | " | 13-Aug | At Cape Mele | 2541 |
| Steamer Ivar | Danish | 13-Aug | Cardiff-Genoa route | 2139 |
| Sailing Vessel Francesco Saverio | Italian | 13-Aug | 15 miles from Noli | 214 |
| Sailing Vessel Emilia | " | 14-Aug | 10 miless off Cape Corvo | 319 |
| | | | Continued next | 81393 |

| Vessel class and name | Flag | Month and day | Location where Sunk | Tons |
|---|---|---|---|---|
| | | | previous page | 81393 |
| Sailing Vessel Rosario | " | 14-Aug | Off Bastia | 188 |
| Sailing Vessel Henriette B. | " | 14-Aug | 10 miles from Capraia Island | 176 |
| Sailing Vessel Francesca | " | 14-Aug | 8 miles from Cape Cosre | 161 |
| Sailing Vessel Lavinia | " | 14-Aug | 12 miles from Capraia Island | 243 |
| Steamer Giovanni Battista | " | 14-Aug | Off Bastia | 1067 |
| Sailing Vessel Pausania | " | 14-Aug | Northern tip of Corsica | 107 |
| Sailing Vessel Louis B. | " | 14-Aug | 15 miles SE from Capraia Island | 212 |
| Sailing Vessel Ida | " | 14-Aug | 9 miles SE from Capraia Island | 242 |
| Sailing Vessel San Guiseppe Patriarca | " | 14-Aug | North of Capraia Island | 62 |
| Sailing Vessel Francesco di Paoio S | " | 14-Aug | Northern tip of Corsica | 112 |
| Steamer Augusta | " | 15-Aug | 45 miles from Cape Figari | 523 |
| Sailing Vessel Virginia di Pompei | " | 15-Aug | 41° 45' N, 10° 2' E | 145 |
| Sailing Vessel Candida Altieri | " | 15-Aug | 25 miles NW from Ponza | 282 |
| Sailing Vessel Madre | " | 16-Aug | 55 miles NNW from Capraia | 665 |
| Steamer Swedish Prince | English | 17-Aug | 12 miles NW from Pantellaria | 3712 |
| Steamer Erix | Italian | 18-Aug | At Cape Panero | 923 |
| | | | | 90213 |

| Vessel class and name | Flag | Month and day | Location where Sunk | Tons |
|---|---|---|---|---|
| **U 35 - von Arnauld - September 14 to October 9, 1916.** | | | | |
| Sailing Vessel Doride | Italian | 19-Sep | 70 miles W of Marittimo | 1250 |
| Sailing Vessel Teresa C. | " | 19-Sep | 53 miles SSE from Cape Carbonara | 270 |
| Sailing Vessel Garibaldi | " | 22-Sep | 110 miles from Minorca | 1373 |
| Steamer Giovanni Zambelli | " | 22-Sep | South of the Balearic Islands | 2485 |
| Steamer Charterhouse | English | 23-Sep | 26 miles ES1/2S from Formentera | 3021 |
| Steamer Nicolo | Italian | 24-Sep | 19 miles N of Dragonera Island | 5466 |
| Steamer Bufjord | Norway | 24-Sep | In the open sea | 2284 |
| Steamer Bronwen | English | 24-Sep | 25 miles NE from Dragonera | 4250 |
| Steamer Benpark (ex Liverpool) | Italian | 25-Sep | 40 miles SE from Barcelona | 3842 |
| Steamer Stathe (ex-Tregenna) | English | 26-Sep | 50 miles ES from Barcelona | 2623 |
| Steamer Newby West | " | 26-Sep | 53 miles ES from Barcelona | 2168 |
| Steamer Boddam | " | 26-Sep | 76 miles ESE from Barcelona | 3218 |
| Steamer Rullus | " | 27-Sep | 45 miles NEN from Dragonera | 1752 |
| Steamer Secondo | " | 27-Sep | 40 miles NNE from Dragonera | 3912 |
| Steamer Vindeggen<br>? | Norway<br>? | 27-Sep<br>28-Sep | In the open sea<br>SE of Ibiza | 2610<br>? |
| | | | Continued next | 40524 |

May 1916 to February 1917

| Vessel class and name | Flag | Month and day | Location where Sunk | Tons |
|---|---|---|---|---|
| | | | previous page | 40524 |
| Steamer Venus | Italian | 29-Sep | 20 miles from Cape St. Antonio | 3976 |
| Steamer Samos | Greek | 3-Oct | SE of the Balearic Islands | 1186 |
| Steamer Birk | Norway | 4-Oct | SW from St. Antioco | 725 |
| Auxiliary Cruiser Gallia | French | 4-Oct | " | 15045 |
| Steamer Aurora | Italian | 5-Oct | In the open sea | 2806 |
| Steamer Vera (ex Gefle) | Swedish | 5-Oct | SE from Cape Spartivento | 2485 |
| | | | | 66747 |

**U 35- von Arnauld - January 1 to 13, 1917.**

| | | | | |
|---|---|---|---|---|
| Steamer Lesbian | English | 5-Jan | 125 miles ES from Malta | 2555 |
| Sailing Vessel Salvatorc Patre | Italian | 5-Jan | SW from Malta | 200 |
| Steamer Hudworth | English | 6-Jan | 94 miles ESE from Malta | 3966 |
| Steamer Mohacs Field | " | 7-Jan | 40 miles SE from Malta | 3678 |
| Steamer Andoni | " | 8-Jan | 46 miles SE from Malta | 3188 |
| Steamer Lynfield | " | 8-Jan | 32 miles SE from Malta | 3023 |
| | | | | 16610 |

**U 38 - Valentiner - May 12 to 23, 1916.**

| | | | | |
|---|---|---|---|---|
| Steamer Cifford | English | 16-May | South of Crete | 487 |

**U 38 - Valentiner - May 31 to June 14, 1916.**

| | | | | |
|---|---|---|---|---|
| Transport No. 21 | Russian | 8-Jun | Mokopse estuary | 1231 |
| Transport No. 39 | " | 8-Jun | " | 3000 |
| Transport No. 77 | " | 8-Jun | Psezuape estuary | 3000 |
| Sailing Vessel ? | " | 8-Jun | " | 70 |
| Steamer Orion | " | 10-Jun | Near Gagra | 429 |
| | | | | 7730 |

| Vessel class and name | Flag | Month and day | Location where Sunk | Tons |
|---|---|---|---|---|
| **U 38 - Valentiner - June 26 to July 13, 1916.** | | | | |
| Steamer Rockliffe | English | 2-Jun | South of Socha Buitkh Point | 3073 |
| Transport ? | Russian | 8-Jun | North of Hopa | - |
| Transport No. 72 | " | 10-Jun | West of Sukhumi | 3000 |
| | | | | 6073 |
| **U 38 - Valentiner - August 11 to September 7, 1916.** | | | | |
| Transport Remenbrance | English | 14-Aug | In the open sea | 3660 |
| Sailing Vessel Dea | Italian | 19-Aug | 40 miles from San Remo | 166 |
| Sailing Vessel Klios | " | 23-Aug | 17 miles From Cape Faro Punta de Antileo | 190 |
| Sailing Vessel Maria Brizzolari | " | 23-Aug | 50 miles from Cap de la Garoupe | 152 |
| Sailing Vessel Tonina | " | 23-Aug | 70 miles from Dramonte | 137 |
| Steamer Isdalcn | Norway | 24-Aug | In the open sea | 2275 |
| Steamer Iiegoise (ex Tyr) | Belgian | 24-Aug | 75 miles S of Marseille | 3889 |
| Steamer Nostra Signora del Carinina | Italian | 25-Aug | 40 miles E of Cape San Sebastian | 1575 |
| Steamer Leandros | Greek | 25-Aug | 20 miles E of Cape San Sebastian | 1658 |
| Steamer Atlantico | Italian | 26-Aug | 15 miles SE of Formentera Island | 3069 |
| Collier? | English | 26-Aug | At Cape San Sebastian | ? |
| | | | Next page | 16771 |

May 1916 to February 1917

| Vessel class and name | Flag | Month and day | Location where Sunk | Tons |
|---|---|---|---|---|
| | | | previous page | 16771 |
| Steamer Antigoon | Belgian | 29-Aug | Westcoast of Mallorca | 1884 |
| Sailing Vessel Stella del Mare | Italian | 29-Aug | Dragonera Island | 1166 |
| Sailing Vessel Francois Joseph | French | 29-Aug | Westcoast of Mallorca | 114 |
| Steamer Ortrud | Italian | 30-Aug | 30 miles S of Formentera | 1508 |
| Steamer Francesco Musner | " | 31-Aug | 65 miles NW from Algeria | 3225 |
| Steamer Bacchus | French | 31-Aug | 65 miles NW from Algeria | 3538 |
| Steamer Duart | English | 31-Aug | 60 miles N3/4E from Algeria | 3108 |
| Steamer Swift Wings | " | 1-Sep | 18 miles E from Cape Bengut | 4465 |
| Sailing Vessel San Francesca | Italian | 1-Sep | Off the coast of Algeria | 68 |
| Sailing Vessel Uranie | French | 2-Sep | Philippeville | 109 |
| Steamer Strathallan | English | 2-Sep | 20 miles NE of Philppeville | 4404 |
| Sailboat Villadoro | Italian | 3-Sep | 60 miles from Cani Island | 134 |
| Steamer Laristan | English | 4-Sep | 30 miles W from Gozo | 3675 |
| Steamer Saint Marc | French | 5-Sep | East of Malta Channel | 5818 |
| | | | | 49987 |

| Vessel class and name | Flag | Month and day | Location where Sunk | Tons |
|---|---|---|---|---|
| **U 38 - Valentiner - November 17 to December 23, 1916.** | | | | |
| Steamer Michael | Greek | 25-Nov | North from Cape Tenes | 2410 |
| Steamer Chemung | USA | 26-Nov | At Cape Gata | 2615 |
| Submarine Transport Kanguroo | French | 3-Dec | At Madera Island | 2493 |
| Cable Steamer Dacia | English | 3-Dec | Funchal Raid | 1856 |
| Steamer Brask | Norway | 8-Dec | In the atlantic ocean | 1464 |
| Steamer Britannia | English | 8-Dec | 70 miles WS from Cape Sines | 1814 |
| Sailing Vessel Brizela | Portuguese | 9-Dec | 70 miles WS from Cape Sines | 200 |
| Steamer Angelo Parodi | Italian | 13-Dec | 6 miles from Cape Tingoso | 3825 |
| Steamer Esemplare | " | 10-Dec | 55 miles from Cape S. Maria | 2595 |
| Steamer Kaupanger | Norway | 13-Dec | 37° 2' N, 0° 48' W | 3354 |
| Steamer Emanuele Assapa | Italian | 15-Dec | 3 miles from Cape Bengut | 3442 |
| Sailing Vessel Tripoli | " | 17-Dec | SE from Cape Carbonara | 56 |
| Steamer Itonus | English | 20-Dec | 60 miles NW-W1/2W from Malta | 5340 |
| | | | | 31464 |
| **U 38 - Valentiner - January 21 to February 18, 1917.** | | | | |
| Steamer Monte Ventoux | French | 25-Jan | Malta-Cerigo Route | 3223 |

| Vessel class and name | Flag | Month and day | Location where Sunk | Tons |
|---|---|---|---|---|
| **U 39 - Forstmann- May 16 to June 6, 1916.** | | | | |
| Steamer Baron Tweedmouth | English | 30-May | 25 miles NEN from Cape Carbon | 5007 |
| Steamer Salmonpool | English | 1-Jun | 30 miles NEE from Cape Carbon | 4905 |
| Stamer Lady Ninian | " | 28-May | 106 miles NE1/2N from Algeria | 4297 |
| Steamer Elmgrove | " | 29-May | 96 miles NE from Algeria | 3018 |
| Steamer Trunkby | " | 2-May | 50 miles SE from Port Mahon | 2635 |
| Southgarth | " | 29-May | 60 miles NNE from Algeria | 2414 |
| Steamer Dalegarth | " | 30-May | 12 miles NE from Cape Corbelin | 2265 |
| Steamer Derwsland | " | 1-Jun | 28 miles NEE from Cape Carbon | 1993 |
| Steamer Baron Vernon | English | 29-May | 56 miles NE1/2N from Algiers | 1779 |
| Steamer Mar Terso | Italian | 27-May | 65 miles SE from Port Mahon | 3778 |
| Steamer Moravia | " | 25-May | Southcoast of France | 3506 |
| Steamer Washington | " | 23-May | Near Piombino | 2819 |
| Steamer Hermersberg | " | 30-May | African northcoast | 2824 |
| Steamer Hercules | " | 23-May | 70 miles SE from Genoa | 2704 |
| | | | Next page | 43944 |

| Vessel class and name | Flag | Month and day | Location where Sunk | Tons |
|---|---|---|---|---|
| | | | From previous page | 43944 |
| Steamer Birmania | " | 21-May | NE from Marittimo | 2215 |
| Steamer Rauma | Norway | 30-May | NE from Cape Carbon | 3048 |
| Steamer Aurrera | Spanish | 24-May | In the Gulf of Genoa | 2845 |
| Sailing Vessel Rosalie Madre | Italian | 21-May | At Cape Carbonaro | 250 |
| Sailing Vessel Valsesia | " | 20-May | 15 miles from Cape Bion | 247 |
| Sailing Vessel Redentore | " | 20-May | 7 miles from Rosrello | 227 |
| Sailing Vessel Rita | " | 25-May | 12 miles S of Porquerelles Island | 200 |
| | | | | 52976 |

**U 39 - Forstmann- Juy 5 to August 1, 1916.**

| Vessel class and name | Flag | Month and day | Location where Sunk | Tons |
|---|---|---|---|---|
| Steamer Ville de Rouen | French | 13-Jul | In the open sea | 4721 |
| Steamer Ecclesia | English | 14-Jul | 11 miles NW from Cape Bougaroni | 3714 |
| Steamer Antigua | " | 14-Jul | 20 miles E-E from Djidjelli (Algierian coast) | 2876 |
| Steamer Sylvie | " | 15-Jul | 15 miles from Cape Sigli | 1302 |
| Steamer Euphorbia | " | 16-Jul | 56 miles NE from Algeria | 3837 |
| Steamer Sirra | Italian | 16-Jul | 60 miles from Cape Kassino | 3203 |
| Steamer Wiltonhall | English | 16-Jul | 65 miles NW from Algiers | 3387 |
| | | | Next page | 23040 |

May 1916 to February 1917

| Vessel class and name | Flag | Month and day | Location where Sunk | Tons |
|---|---|---|---|---|
| | | | From previous page | 23040 |
| Steamer Angelo | Italian | 17-Jul | 80 miles NE from Algiers | 3609 |
| Steamer Roosemoor | English | 17-Jul | 80 miles NE-E from Algiers | 4303 |
| Steamer Llongwen | " | 18-Jul | 90 miles NE from Algiers | 4683 |
| Steamer Grangemoor | " | 20-Jul | 75 miles NW-W from Algiers | 3198 |
| Steamer Yser | " | 20-Jul | 56 miles NW1/2N from Algiers | 3538 |
| Steamer Cettois | French | 20-Jul | In the open sea | 948 |
| Steamer Karmaa | English | 20-Jul | 68 miles NNW from Algiers | 3710 |
| Steamer Wolff | " | 21-Jul | 75 miles NNW from Algiers | 2443 |
| Steamer Knutsford | English | 22-Jul | 12 miles NWN of Cape Corbelin | 3842 |
| Steamer Olive | " | 22-Jul | 1 miles NW from Cape Corbelin | 3678 |
| Steamer Badminton | " | 23-Jul | 63 miles NEE from Cape Carbon | 3847 |
| Sailing Vessel Maria | Italian | 24-Jul | At Elba Island | 198 |
| Steamer Letimbro | " | 29-Jul | 40 miles from Benghazi | 2210 |
| Sailing Vessel Rosarine | " | 29-Jul | 30° 40' N. 19° 5' E | 131 |
| | | | | 63378 |

| Vessel class and name | Flag | Month and day | Location where Sunk | Tons |
|---|---|---|---|---|
| **U 39 - Forstmann- October 8 to November 1, 1916.** | | | | |
| Steamer Penylan | English | 19-Oct | 5 miles W-N of Cape Bougaroni | 3875 |
| Steamer Mombassa | " | 20-Oct | 8 miles NW-N of Cape Carbolin | 4689 |
| Steamer Ciuden | " | 22-Oct | 11 miles West from Cape Tenes | 3166 |
| Steamer W.Harkess | English | 22-Oct | 17 miles West from Cape Tenes | 1185 |
| Steamer Nina | Italian | 22-Oct | 10 miles West from Cape Tenes | 3383 |
| Steamer Ravn | Norway | 22-Oct | In the open sea | 998 |
| | | | | 17296 |
| **U 39 - Forstmann- November 18 to December 10, 1916.** | | | | |
| Steamer Margarita | Greek | 27-Nov | NW from Alexandria | 1112 |
| Steamer Reapwell | English | 27-Nov | 148 miles NW-N from Alexandria | 3417 |
| Steamer King Malcolm | " | 28-Nov | 144 miles NW-N from Alexandria | 4351 |
| Transport Moresby | " | 28-Nov | 120 miles NW-N from Alexandria | 1763 |
| Steamert Istrar | " | 2-Dec | 120 miles NNW1/2W from Alexandria | 4582 |
| Transport Plater | " | 3-Dec | NW from Alexandria | 4243 |
| | | | | 19468 |

| Vessel class and name | Flag | Month and day | Location where Sunk | Tons |
|---|---|---|---|---|
| **U 39 - Forstmann- January 3 to 27, 1917.** | | | | |
| Steamer Baynesk | English | 9-Jan | 130 miles NW from Alexandria | 3286 |
| Oil tanker Garfield | English | 15-Jan | 16 miles NE-N1/2N from Alexandria | 3838 |
| Transport Amiral Magon | French | 25-Jan | In the open sea | 5566 |
| | | | | 12690 |
| **U 63 - Otto Shultze - November 23 to Deceomber 14, 1916.** | | | | |
| Steamer Maude Larssen | English | 27-Nov | 22 mile WSW from Marittimo | 1222 |
| Steamer Sigurd | Danish | 28-Nov | 15 miles SW of Cape Spartivento | 2119 |
| Sailboat Roma | Italian | 30-Nov | At Gross Island | 642 |
| Sailing Vessel Luigi C. | " | 2-Dec | " | 71 |
| Steamer Grigorius Angelatos | Greek | 5-Dec | In the open sea | 3632 |
| Steamer Magellan | French | 11-Dec | SE from Pantellaria | 6027 |
| Steamer Sinai | " | 11-Dec | In the open sea | 4623 |
| | | | | 18336 |
| **U64 - Moraht - October 26 to November 19, 1916.** | | | | |
| Steamer Tripel | Norway | 10-Nov | 3.5 miles from La Carrucha | 4633 |
| Steamer Freya | Danish | 10-Nov | 10 miles East from La Carrucha | 2168 |
| Steamer Matarazzo | English | 15-Nov | In the open sea | 2823 |
| | | | | 9624 |
| **U 65 - von Fischel - December 2 to 6 , 1916.** | | | | |
| Steamer Caledonia | English | 4-Dec | 125 miles E-S from Malta | 9223 |

| Vessel class and name | Flag | Month and day | Location where Sunk | Tons |
|---|---|---|---|---|
| **U 72 -Kraft - November 16 to Deember 17, 1916r.** | | | | |
| Sailing Vessel Maria di Pompei | Italian | 19-Nov | 115 miles South of Cape Passero | 286 |
| Sailing Vessel Marguerita F | " | 23-Nov | At Cape Figari | 44 |
| Steamer Christoforus | Greek | 26-Nov | In the open sea | 3674 |
| Sailing Vessel Salvatore Ciampa | Italian | 27-Nov | " | 1728 |
| Steamer Palermo | " | 2-Dec | At Cape San. Sebastian | 9203 |
| Sailing Vessel Jeanne | " | 2-Dec | La Calella | 534 |
| | | | | 15469 |
| **U 73 - Sieß- July 24 to August 14, 1916.** | | | | |
| Minesweeper Clacton | English | 3-Aug | In the Aegean | 820 |
| SailIng Vessel Lorenzo Donato | Italian | 10-Aug | 65 miles West from Cefalonia | 140 |
| | | | | 960 |
| **U 73 - Sieß- Deceomber 10 to 30, 1916.** | | | | |
| Oil Tanker Murex | English | 21-Dec | 94 miles NW from Port Said | 3564 |
| **UB 42 - Wernicke - August 29 to November 6, 1916.** | | | | |
| Transport Peter Darch | Russian | 3-Sep | In the Black Sea | 731 |
| Sailing Vessel ? | " | 5-Oct | North of Fidonisi | 150 |
| Steamer Czartiza | " | 6-Nov | North of Constanza | 2890 |
| | | | | 3771 |
| **UB 43 - Mellenthin - September 9 to 20, 1916.** | | | | |
| Steamer Italiana | English | 14-Sep | 112 miles North of Malta | 2663 |
| Steamer Lord Tredgar | " | 17-Sep | 51 miles SE-E from Malta | 3856 |
| Steamer Deva | " | 17-Sep | 45° E, 3/4 N off Malta | 3802 |
| | | | | 10321 |

| Vessel class and name | Flag | Month and day | Location where Sunk | Tons |
|---|---|---|---|---|
| **UB 43 - Mellenthin - October 5 to 16, 1916.** | | | | |
| Steamer Welsh Prince | English | 13-Oct | 33 miles SW of Cape Matapan | 4934 |
| Steamer Elax | " | 10-Oct | 70 miles WSW of Cape Matapan | 3980 |
| | | | | 8914 |
| **UB 43 - Mellenthin - October 3 to November 15, 1916.** | | | | |
| Steamer Statesmann | English | 4-Nov | 200 miles East of Malta | 6153 |
| Steamer Clan Leslie | " | 5-Nov | 200 miles E1/2S of Malta | 3937 |
| Steamer Huntsvale | " | 5-Nov | 200 miles East of Malta | 5398 |
| Transport Arabia | " | 6-Nov | 112 miles from Cape Matapan | 7933 |
| Steamer Kahunsa | " | 12-Nov | 205 miles ESE from Malta | 3383 |
| | | | | 26804 |
| **UB 43 - Mellenthin - December 9 to 17, 1916.** | | | | |
| Steamer Bretwalda | English | 13-Dec | 220 miles E-S from Malta | 4037 |
| Steamer Westminster | " | 14-Dec | 196 miles E-S from Malta | 4342 |
| Steamer Russian | " | 14-Dec | 210 miles E-S from Malta | 8825 |
| | | | | 17204 |
| **UB 44 - Wäger - June 20 to July 21, 1916.** | | | | |
| Steamer Moeris | English | 30-Jun | 46 miles SE from Sidero | 3409 |
| **UB 45 - Palis - July 13 to 21, 1916.** | | | | |
| Steamer Virginia | English | 16-Jul | 42 miles from Cape Matapan | 4279 |
| Steamer Anglesea | " | 18-Jul | Escaped | - |
| | | | | 4279 |

| Vessel class and name | Flag | Month and day | Location where Sunk | Tons |
|---|---|---|---|---|
| **UB 45 - Palis - August 24 to September 6, 1916.** | | | | |
| Steamer Tewere | Italian | 31-Aug | Poti (Black Sea) | 2660 |
| **UB 46 - Bauer - July 19 to August 12, 1916.** | | | | |
| Steamer Kohina Maru | Japan | 2-Aug | In the open sea | 3164 |
| Sailing Vessel Bariledos | Greek | 8-Aug | " | 488 |
| | | | | 3652 |
| **UB 46 - Bauer - September 23 to October 7, 1916.** | | | | |
| Steamer Huntsfall | English | 2-Oct | 12 miles SSE from Skyro | 4331 |
| **UB 46 - Bauer - October 29 to November 11, 1916.** | | | | |
| Sailing Vessel Melanie | Russian | 10-Nov | In the Black Sea | 116 |
| **UB 47 - Steinbauer - August 5 to 25, 1916.** | | | | |
| Steamer Stampalia | Italian | 17-Aug | 36° 40' N, 22° 10' E | 8999 |
| Steamer Pacuare | English | 19-Aug | Escaped (3897t) | - |
| | | | | 8999 |
| **UB 47 - Steinbauer - September 3 to 17, 1916.** | | | | |
| Steamer Llangorse | English | 8-Sep | 48 miles WSW of Cape Matapan | 3841 |
| Transport Butetown | " | 8-Sep | 55 miles WSW of Cape Matapan | 3789 |
| Steamer Spetzai | Greek | 8-Sep | In the open sea | 1904 |
| | | | | 9534 |

May 1916 to February 1917

| Vessel class and name | Flag | Month and day | Location where Sunk | Tons |
|---|---|---|---|---|
| **UB 47 - Steinbauer - September 27 to October 18, 1916.** | | | | |
| Steamer Franconia | English | 4-Oct | 195 miles E1/2S from Malta | 18150 |
| Transport Crosshill | " | 11-Oct | 60 miles West from Malta | 5002 |
| Steamer Avis | Greek | 15-Oct | In front of Messina | 1000 |
| Sailing Vessel Annunziato | Italian | 14-Oct | At Syracuse | 61 |
| Sailing Vessel Retendore | " | 14-Oct | " | 79 |
| Sailing Vessel Elena | " | 14-Oct | " | 52 |
| Sailing Vessel Dundromi | " | 14-Oct | " | 50 |
| Sailing Vessel La Nuova Concettina | " | 14-Oct | " | 32 |
| | | | | 24426 |
| **UB 47 - Steinbauer - December 17 to January 7, 1917.** | | | | |
| Transport Ivernia | English | 1-Jan | 58 miles SE1/4E of Cape Matapan | 14278 |
| Steamer? | - | - | Missing data | - |
| | | | | 14278 |
| **UC 22 - von Heimburg - September 22 to October 12, 1916.** | | | | |
| Sailing Vessel Emma | Russian | 28-Sep | English West Coast | 279 |
| **UC 22 - von Heimburg - November 25 to December 7, 1916.** | | | | |
| Steamer Luciston | English | 29-Nov | 4 miles E of Delamara Strait Malta | 2948 |
| Steamer Burcombe | " | 1-Dec | 100 miles SE1/2E of Malta | 3516 |
| Steamer Algerie | French | 4-Dec | 120 miles East from Malta | 4034 |
| | | | | 10498 |

| Vessel class and name | Flag | Month and day | Location where Sunk | Tons |
|---|---|---|---|---|
| **UC 22 - von Heimburg - December 23 to January 5, 1917.** | | | | |
| Steamboat Oronsay | English | 28-Dec | 48 miles SE from Malta | 3761 |
| Steamer Apsleyhall | " | 30-Dec | 28 miles W-N from Gozo | 3882 |
| Steamer Baycraig | " | 1-Jan | 84 miles ESE from Malta | 3761 |
| | | | | 11404 |
| **UC-37. From 23 / XII 1916 to 12/1 1917 (Transition from the North Sea to the Mediterranean)** | | | | |
| Steamer Britannic | Norway | 1-Jan | Off the coast of Portugal | 2289 |
| Steamer Fama | " | 2-Jan | " | 2417 |
| Steamer Dimitrlos Goulandris | Greek | 2-Jan | " | 3744 |
| Steamer Aristoteles C. Joannon | " | 2-Jan | " | 2868 |
| Sailing Vessel Notre Dame | French | 3-Jan | " | 267 |
| Sailing Vessel de Berger Capricienne | " | 3-Jan | " | 159 |
| Sailing Vessel Liberte | " | 3-Jan | " | 250 |
| Steamer Luigi Stampa | Italian | 4-Jan | " | 3987 |
| Sailing Vessel Secmel | Russian | 4-Jan | " | 209 |
| Steamer Wragby | English | 4-Jan | 45 mile W-N of Cape Spartel | 3641 |
| Sailing Vessel Doribe | Italian | 5-Jan | East of Gibraltar | 1250 |
| | | | | 21081 |

## Baltic Sea

A relatively large number of submarines operated in the Baltic Sea combat theater in the period in question. The stagnation of trade warfare in the North Sea caused the Naval Staff to transfer boats “U19", “U22” and “U47"from there to the Baltic Sea. Three large, five (UB) small torpedo submarines and three (UC) mine submarines (see compilation following pages) were used there during this time.

As in previous sections, their operations supported raids into the Gulf of Bothnia and Gulf of Finland as well as waiting positions in the area of the Aland Islands.

Merchant ship sinkings achieved and where UC-boats carried out minelaying, is apparent from the attached tables. (two tables in the following pages.)

Particularly noteworthy is an operation made by “UB30”, “UB19” and “UB33” in the Gulf of Riga (September 11-17, 1916). All three boats sighted Russian cruisers and many destroyers; "UB31" (Oblt. Bieber) came closest to a success, the boat was about to attack an enemy cruiser and was finally forced away at the last moment by a destroyer. Much like in the Black Sea, Russians ships in this theater of war maneuvered extremely skillfully against submarines. Like our other enemies, Russians also began to use depth charges against our boats and airplane bombs.

**Mine Operations in the Baltic Sea. May 1916 to January 1917.**

| Boat | Cdr. Name | Sortie Date | Minefield location | Remarks |
|---|---|---|---|---|
| UC 4 | Oblt. Hamm | May 30 to June 2 | Western exit of the Gulf of Riga | - |
| | Oblt. Pilzecker | August 18 to 25 | North entrance to Moon Sound. | - |
| | | August 30 to Septemb er 3 | Between Hango Utö - Abo | Sept. 1 commenced return due to commander' s illness |
| UC 25 | KptLt. Feldkirchner | Septemb er 28 to 30 | Between Fästorne Bank and Lagskär | - |
| | | October 14 to 22* | Eastern part of the Gulf of Finland | October 21 sighted enemy submarine. No attack possible |
| | | Novembe r 14 to 17 | Exit to Soelo Sound | - |
| | | Novembe r 21 to 24 | In front of Soelo Sound | - |
| UC 27 | Oblt. Vesper | Septemb er 22 to 26 | before Zerel | - |
| | | October 1 to 7 | Odensholm — Spithamn; Gras Grund — Great Rogö; Packerort Bank; at Surop | October 6 trapped in a net |

| Boat | Cdr. Name | Sortie Date | Minefield location | Remarks |
|---|---|---|---|---|
| | | October 22 to 30 | North of Highlands Island; Groß-Tjuters Light. – Wikalla South Buoy; at Wulf Island** | - |
| | | November 8 to 16 | in front of the Led Sound, near Hammarudde; Niemi-Santakari; Walki Akari; west of Pitterskärz, southwest of Kallo | - |
| | | November 27 to December 2 | - | December 1 returned due to compass malfunction |

The Russian language version includes the following comments: * Barrier at about Lavensaari was discovered thanks to a floating mine. ** The following were blown up at these barriers: a) at the South Gogland lighthouse - the armored cruiser Rurik (September 19, 1916[Julian calendar]); b) The destroyer Letun at Wulf Island.

## Ships sunk in the Baltic Sea. May 1916 to January 1917

| Vessel class and name | Flag | Month and day | Where Sunk | Tonnage |
|---|---|---|---|---|
| **U 19 - Spieß - September 9 to 26, 1916** | | | | |
| Transport unknown* | unknown | Sep-20 | Gulf of Finland | ? |
| Transport No. 18 | " | Sep-20 | " | ? |
| Steamer Kennett | English | Sep-20 | " | 1679 |
| | | | | 1679 |
| **U 22 - Scherb, October 30 to November 15, 1916** | | | | |
| Sailing vessel Fanadis | Russian | 1-Nov | Gulf of Bothnia | 384 |
| Coaster Frans | Swedish | 2-Nov | Sweden - Finland | 133 |
| Motorsailer Jonköping | " | 3-Nov | " | 81 |
| Steamer Agir | " | 3-Nov | " | 427 |
| Motorsailer Taimi | Finnish | 8-Nov | Gulf of Bothina | 114 |
| Steamer Astrid | Swedish | 11-Nov | Sweden - Finland | 191 |
| Steamer Rundild | " | 2-Nov | | 1380 |
| | | | | 2710 |
| **U47 - Metzger, August 16 to September 3, 1916** | | | | |
| Steamer Wellamo | Russian | 29-Aug | Gulf of Bothnia | 1050 |
| Steamer Presto | Swedish | 29-Aug | Brought to Libau as a prize | - |
| | | | | 1050 |
| **UB 20 - Viebeg - July 28 to August 6, 1916** | | | | |
| Steamer Bror Oscar | Swedish | 2-Aug | Gulf of Bothnia | 368 |
| Steamer Varmland | " | 2-Aug | " | 213 |
| Brig Vera | " | 2-Aug | " | 312 |
| Steamer Commerce | " | 3-Aug | " | 368 |
| Steamer Temis | " | 2-Aug | Brought in as a prize | - |
| | | | | 1531 |

| Vessel class and name | Flag | Month and day | Where Sunk | Tonnage |
|---|---|---|---|---|
| **UB 30 - Montigny - October 18 to 28, 1916** | | | | |
| Bark August | Swedish | 21-Oct | at Finngrundet Lightship | 346 |
| Motorsailor Elly | " | 23-Oct | NW from Mäntyluoto | 88 |
| | | | | 434 |
| **UB 36 - Albrecht - July 27 to August 4, 1916** | | | | |
| Steamer Hudiksvall | Swedish | 2-Aug | Baltic Sea | 481 |
| Steamer Peter Brake | Finnish | 2-Aug | " | 499 |
| Sailing vessel Anna | Swedish | 30-Jul | " | 172 |
| Steamer Pita | " | 30-Jul | Brought in as a prize | - |
| | | | | 1152 |

[Editor: The Russian language version includes the following comments: * "U19" on September 12 attacked a detachment of seven transports near the Baltic port. Hit by two torpedoes, The transport "Elizabeth" ran ashore. In addition, "UB81" Vesper, Bieber and "UB33" Oblt. Lefholz worked in the Baltic Sea.]

## Conclusion

In retrospect, it should be noted once again submarine warfare in the period from May 1916 to the end of January 1917 generated successes, per boat and per cruise, in all operational areas outside the enemy's closely guarded zones, which were not later equaled by any type of submarine warfare, even unrestricted warfare; because our opponents, led by Britain, feverishly developed anti-submarine countermeasures. The full extent of this possibility was exploited only in the Mediterranean Sea. It was particularly useful when Flanders sent the newer BII boats on its own initiative to the western English Channel commencing in September 1916 to conduct trade warfare with Prize Orders, as did the constant striving of the KKpt. Bartenbach to gradually increase the activity of his boats. Abandoning trade warfare from May to early October 1916 by the North Sea submarines greatly influenced the outcome of the submarine war.

In the period described Hindenburg and Ludendorff had been given the Supreme Army Command by the Emperor. The two commanders recognized unrestricted submarine warfare was indispensable in order to win the war or in making any decisions thereof. As early as August 30, discussions were held in Pleß with the Chief of the Naval Staff regarding resuming submarine warfare in Flanders and the North Sea. Chancellor Bethmann Hollweg believed a declaration of an unrestricted submarine warfare would mean the entry of Holland and Denmark into the war. Although this fear was somewhat far-fetched (Our naval attache in the Hague at the time, Kapitän v. Müller, and others expressly denied this danger.), the two military leaders believed commencing unrestricted submarine warfare had to be postponed until the war in Romania allowed security forces to be deployed on the Dutch and Danish borders. In any case, the decision in favor of unrestricted submarine war was hereby made in principle and only as yet unclear regarding timing. It should therefore be noted the decision to launch a large submarine construction program was made at this time.

Until now, generous orders for large submarine series had been deliberately omitted because the choice of submarine types to be commissioned depended on changing operational use of submarines and the individual submarine types newly constructed during the war initially needed to prove themselves at the front before they were put into construction in large numbers. Now these obstacles no longer existed. Large submarines series could be ordered. This is the only manner to achieve full technical utilization of submarine yards and the industry that supplies them.

After the Pleß meeting in 1916 only two 800 ton boats, sixteen 400 ton boats and nine cruiser submarines were ordered.[198] Anticipating somewhat it should be mentioned the February 1917 order for six 800 ton boats, 45 400 ton boats, did not match the significance of the declaration of unrestricted submarine warfare made on February 1, 1917. It was not until June and December 1917 that larger submarine orders were placed with the shipyards. This will be discussed in the 5th volume. The order for nine submarine cruisers, (U-cruisers), in 1916 was absolutely correct. A careful study of the operations against trade led to the logical conclusion of the need for new operational methods to achieve decisive successes to counter the weight of growing enemy opposition. The most important of these new methods was to carry the submarine trade war into the vastness of the Atlantic Ocean and to relaunch the operations of the "Karlsruhe" and the auxiliary cruiser "Möwe". This operation - e.g. To commence at the end of 1917 with ten to twelve new cruiser submarines - would have encountered few defenses on the east coast of South America, from which considerable English supplies originated, and the introduction of the defensive measures in this region would naturally have thinned those around England. Subsequently, the greatest difficulties for the small boats were associated with the need of frequent passage through hazardous zones off their own coast. A submarine cruiser that passed through these zones much less often because it could stay at sea for four months was much less at risk. The six commercial submarines that remained after the loss of the "Bremen" were converted into submarine cruisers in 1917, but they were too slow (7 knots) to

---

[198] In a postwar lecture Gayer adds more information. "Particularly fateful was the decision by Navy Secretary of State v. Capelle, which was influenced by the view that a large number of submarines would be disadvantageous postwar for the administration and expansion of the navy as a world power." The Director of Submarine Construction at the Admiralty, KKpt. Spindler, made urgen representations to order new submarines. This was only acceded to after many delays and refusals. Gayer muses it should have been the other way around, i.e. the secretary should have expressly ordered new constructions with all resources available. The source used is a contemporary translation by Commander W.P. Beehler, USN, titled "Summary of German Submarine Operations in the Various Theaters of War From 1914 to 1918", published in the United States Institute Proceedings vol.52, no.4, whole no. 278, April 1926, p.650.

replace a modern submarine cruiser's 15 knot capability. We will see later submarine cruiser construction was repeatedly postponed in favor of small boat construction, which seemed necessary to compensate for their increasing losses, and so large submarine cruiser operations in the vast Atlantic did not come about. This also affected the outcome of the submarine campaign.

On January 9, 1917, torpedoing armed merchant ships without warning was authorized after lengthy negotiations with the political authorities and the General Staff. On January 12, 1917, unrestricted submarine warfare was announced and its commencement date fixed at February 1, 1917. (See map at end of chapter in Fig 26.[199]) Despite the late date, the decision to proceed with unrestricted submarine war was greeted with joy by submariners at the front and felt to be a release from the eternal back and forth of operations, although now and then there was the dull feeling that it was too late and justified doubts as to whether the entire organization would prove materially equal to the great, finally made decision.

Unfortunately, the unrestricted submarine warfare declaration was accompanied by a Greek gift. It was the Naval Staff's declaration in the spring of 1916 hoping England would be forced to make peace in six months by using such methods of warfare.

This declaration had two consequences.

First of all, it led to a certain overload at the submarine front, if there was reason to believe it would end in six months it was natural to try to shorten the length of time boats were docked between different cruises; complete overhauls were no longer considered necessary seeing as the forthcoming long war seem unlikely. The six-month principle also seemed to justify a tougher use of the submarines. In this respect, the idea of the North Sea boats taking the shortest route through the Dover-Calais passage to stretch their operation areas played a role, even though this passage

---

[199] Gay, George I. Public Relations of The Commission for Relief in Belgium. Documents, Stanford, Stanford University Press. 1929.

has been avoided with few exceptions since April 1915 by the large submarines. However, Flanders boats had meanwhile acquired considerable experience in overcoming the Passage's difficulties and the obstacles newly conceived by the enemy were in a sense under constant surveillance by the Flanders Flotilla. For this reason Captain Bartenbach advised North Sea submarines to use the passage. Since the commencement of unrestricted submarine warfare it was initially utilized in about 50% of trips to the west coast of England. We will see how this operational question developed in the 5th volume.

Second, the success promised by the Naval Staff in such a short time paralyzed the decision to immediately start a generous submarine construction program. It has already been mentioned that construction orders in 1917 did not in any way correspond to the scope of the unrestricted submarine warfare declaration. Admiral v. Capelle explained the early success of the submarine war, as expected by the Naval Staff, prevented him from ordering submarines, the first of which could only be ready a year later.

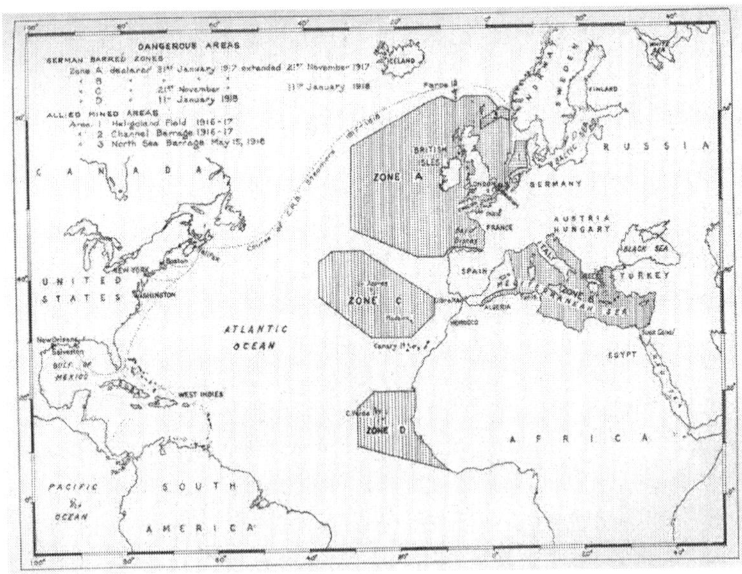


Fig 26. Unrestricted Zones

# Afterword[200]

The initial declaration of unrestricted submarine war stipulated the waters around Great Britain and French coast to the Atlantic Ocean.(extended further west in November 1917).[201] The Dover Strait sortie route for large North Sea based boats proved straightforward for some boats while many other experienced considerable difficulty in the passage due to unfavorable weather and other conditions; the North Sea route being an alternative one. In March, the Commander of North Sea Submarines requested and was denied permission (by the Fleet Commander) to personally attempt an English Channel passage and observe operating conditions via that route. [202] Afterwards all of the large North Sea boats were directed to use the route north about England.

Tonnage sunk immediately increased in 1917 following the declaration of unrestricted submarine warfare. (Fig 27 & 28 at end of text) In February double of what was achieved in January, while in April and June nearly a million tons. Some boat commanders achieved exceptional results; KptLts. Schweiger, Rose, Wunsche, v. George, Adam, and Dichman being singled out. The tonnage sunk in single patrols by KptLt. Petz in "U85", (50,000 tons in Dec.) and Rose (March[203]) being cited.

---

[200] The source used is a contemporary translation by Commander W.P. Beehler, USN, titled "Summary of German Submarine Operations in the Various Theaters of War From 1914 to 1918", published in the United States Institute Proceedings vol.52, no.4, whole no. 278, April 1926, pp.621-659.

[201] Followed in November 1917 by the waters around the Azores and Madeira Islands (expanded January 1918) and then those around the Cape Verde Islands-bulge of Africa in January 1918. around the Azores and also northeast US coastline towards Canadian east coast. See Fig 26.

[202] In 1917 there were approximately 153 Dover Straits passages with two losses. Four submarines were lost attempting the passage between Jan. 19 and Feb. 8, 1918.

The extraordinary success rate and the condition of requiring England to be forced to sue for peace within six months meant construction was delayed as the boats constructed would not be available in the time frame indicated. Although insufficient boat quantity was at the Front in February 1917, Admiral Capelle himself said he could not, under the circumstances, agree to order boat construction which could not be completed within one year's time.

Meanwhile England took the lead in developing countermeasures with alacrity noting the momentous events which were occurring regarding supplying the war effort. Around the middle of 1917 countermeasures were focused on three endeavours:

1. Increased net and minelaying operations near the German Bight.
2. Further obstacles to block the Dover Straits.
3. Institution of merchant convoys.

Germany responded to the first countermeasure by use the Fleet more often to keep egress/ingress routes open and to countermine; the latter achieving some success as indicated by published loss lists.[204] On the debit side several submarines were lost to mines in the Bight. An example being "U59" under KptLt. v. Firks, who struck a mine on May 14, 1917 in the area of Horns Reef, in the fog he missed the navigational markers indicating the route southward. From the middle of 1917 onwards, the commanders of mine-sweeping forces and the fleet commander worked together to keep clear a large area through the Bight, commencing off the Ems northward and west to the outer limits of English minefields.[205] Submarines were escorted through this area and then proceeded submerged through the English outer fields. On entering the reverse procedure was followed. The scheme initially worked well. British submarines observed and reported these methods and the British

---

[203] Rose sank nearly 30,000 tons.

[204] The British paused their mining activity in August 1918 after a destroyer flotilla ran into a German minefield.

[205] Approximately 33 German surface vessels were lost in minesweeping operations in 1917.

changed their mining methods. Instead of fields, a system of nets and mines placed at different depths was utilized. KpLts. Schweiger of "U88", Berger of "U50" and Hufnagel of "U106" were lost in this manner. (The latter on October 9, 1917) Indecision regarding how to permanently deal with this new barrier led to the boats using via the Baltic through the Little Belt.[206]

Commencing in September 1917, the new Commander of Submarines, Commodore Michelsen, elected to utilize the Dover passage for the large North Sea based boats. Some boats had difficulties: On September 28, "U70" was caught by the stern in a mined net and was able to put into Zeebrugge. In November, Both "U48" and "U94" ran aground on the Goodwin Sands due to gyro-compass failure and strong currents. "U94" came clear after lightening ship;"U48" was blown up by her crew. Some commanders had excellent results; KptLt. Rose in "U53" sank 28,000 tons in a thirteen day cruise in October. Commodore Michelsen, resumed using the route north about/around England in January 1918 after the unexplained loss of "U87" (Spetz-Schulzburg), "U84" (Rohr), "U93" (Gerlach) and "U95"(Brinz) and reports from the Flanders Flotilla of enhanced barriers in the Dover Straits.

Convoys proved to be a very effective countermeasure. Conceived at the height of operations in June 1917, the assembly of twenty to thirty merchantmen in convoys considerably reduced opportunities for attacks. Further the attack was made technically much more difficult due to the escorting screen. The convoy system was soon deployed in all war theaters-Atlantic Ocean, Mediterranean, North Sea, and between the Shetlands and Bergen. This latter had such heavy traffic that specific submarine operations were organized to operate against it, initially with large boats, then with BII boats and finally with surface strike forces.

---

[206] Admiral Scheer noted in his memoirs that he barely had sufficient ships to search for mines. Op cited, pp.289-290.

These countermeasures combined effectively prevented any additional sinking above the peak number amassed in June 1917, although an alarming decline did not occur until May 1918. Tonnage could have remained at the June 1917 level if more boats were available. Had a building program been initiated in August 1916, when Hindenburg and Ludendorff elected upon unrestricted submarine warfare as a policy decision, they would have been operational in late 1917. The largest number of boats available for operations at the Front peaked at 132 in September 1917, fluctuating thereafter from 110 in June 1918 to 123 in October 1918. Under these circumstances commanders at the front urged to increase and hasten submarine construction in the fall of 1917. The efforts of both Michelsen and Scheer led to the creation in late 1917 of an independent Admiralty bureau responsible for all material, personnel, and organization of the submarine service. (Previous efforts to establish a bureau had not led to any results.) The moment appeared favorable as it was hoped the collapse of Russia would release additional workers for submarine construction. The Army disagreed, wanting to keep industry focused on land warfare armaments in view of their forthcoming Western Front offensive.

It was debated within the Submarine Bureau which option would yield best results- focusing labor on submarine construction or land warfare armaments. Additionally the manpower released from service in Russia could be used to crush Italy, thereby supporting the efforts of Mediterranean submarines who had been so successful throughout the period. The Chief of the Submarine Bureau, Vice-Admiral Ritter v. Mann, undertook to present the options to the appropriate channels. Unfortunately there was no consensus. At that time, the Admiralty trusted to the Army High Command, (Hindenburg and Ludendorff), to lead military policy. The Western Front land offensive would take primacy for resources. Admiral Ritter v. Mann was able to influence affairs sufficiently to have small drafts of workmen assigned to submarine related matters. He also better organized delivery of materials to minimize construction delays. Further, the Bureau undertook a large scale construction of repair facilities at existing bases to shorten boat refit times. At the beginning of April, large construction projects were initiated at Pola and Cattaro. (The Pyocia docks project at Cattaro completed in late 1918 when shortly afterwards it became necessary to evacuate the base.) In home waters the following projects were executed: expansion of Holm docks at Danzig, creating a base at Lubeck, another at Hamburg (now Deutsche Werks), and dock expansion at Emden. As early as December 1917, thirty large and ninety small boats were launched and through June 1918, sixty-four additional large and 156 smaller boats. Dockyards who had previously not done submarine construction were co-opted for the work. At the war's end some 440 boats were under construction, 137 being large types. In September and October 1918, contracts for 330 boats were placed of which 117 were large boats. These contracts were negotiated in Berlin with industrialist Hugo Stinnes and other heads of inland industries to diversify labor, to some extent relieving the overloaded dockyards with other industries. The Bureau worked furiously due to the neglect of the preceding years and eventually fought a losing fight. The maximum achieved monthly for completed construction was a peak of thirty new submarines completing construction in September 1918.

Submarine cruiser operations pushed into the far West and South Atlantic commencing in September 1917. These were opened by the cruises of "U152", KptLt. Huesel and "U151", KKpt. Kophamel. Both were converted merchant submarines. These cruises lasted three months and extended as far as the Azores sinking over 40,000tons and 30,000tons respectively. These cruises were directed by the Naval Staff and boded well as there were few submarine countermeasures organized in the operational areas. A particularly successful cruise was made by "U155", KKpt. Eckelmann, (January 14 to May 18, 1918) from the Azores to the Straits of Gibraltar defensive perimeter which resulted in over 50,000 tons. Another cruise by "U152", KptLt. Kolbe, (December 23, 1917 to April 19, 1918) to the triangle formed by Portugal, Africa and the Canary Islands netted 30,000 tons. Only two purpose-built submarine cruisers, U-cruisers, came into service: "U139" and "U140". KKpt. Kophamel commanded "U140" on a cruise to America from July 2 to September 20, 1918. This operation resulted in sinking about 30,000 tons.[207] Operations of both U-cruisers were hindered by pressure hull leaks and poor diving stability. None of the thirty-eight U-cruisers with 3000hp Diesel motors were completed prior to the Revolution came into operation. The fast 1200 ton boats, of which "U135"[208] was the first to be commissioned, suffered the same fate. The new large mine boats of the "U117-126" series, carrying up to twenty-four torpedoes and forty-two mines, were partially operational. The cruise of "U117" under KKpt. Dröscher to the North American coast netted some 23,000 tons.

Undoubtedly these three new types of submarines would have achieved great success if the war had lasted longer. British intelligence was aware and concerned regarding these new developments as substantial possibilities existing for expanding the submarine war to areas far beyond current organized counter-measures. Hence the expression used at the time that Germany had given up the struggle "five minutes" too soon.

Major warships continued to be attacked when opportunity offered. One January 9, 1917 "U32", KptLt. Hartwig, sank the English battleship "Cornwallis" near Malta. On March 19, 1917, "U64, KptLt. Morath, sank the French battleship "Danton". On December 14, 1917, the French armored cruiser "Chateaurenault" by "UC38", commanded by KptLt. Wendlandt, in which engagement the latter was also sunk.[209]

In the Mediterranean German submariners from "U35"took a large part in combating the Austrian sailors revolt in Cattaro in February 1918. The collapse of Austria forced Germans submarines to

---

[207]KptLt. von Arnauld de la Perière commanded "U139", (He returned home from "U35" and the Mediterranean after his final March 1918 cruise.), departing Kiel on September 1, 1918 on a cruise to America via the Azores. He sank about 7,000 tons before being recalled home on October 21.

[208] "Medium-U" as the class was known forming "U127-U138".

[209] KptLt.Ernst Hashagen, "U62", sank the French armored cruiser on "Dupetit-Thouars" on August 7, 1918 in the Bay of Biscay.

abandon the Mediterranean for German ports in October 1918. Some of the large and small boats had to be scuttled during the evacuation of Pola and Cattaro as they could not be made seaworthy for the trip home. Passing through the Straits of Gibraltar, "UB50" under Oblt. Kukat sand the British battleship "Russell" with torpedoes near Cape Trafalgar on November 9, 1918. This was the last action at sea of the Mediterranean Flotilla.

The falling off of sinkings from May 1918 onwards was attributed to British anti-submarine measures gaining the upper hand among Flanders boats. As early as January 1918 the average monthly loss rate was four boats. From April onwards the docks at Zeebrugge were subjected to extensive air attacks. The Flanders Flotilla regarded themselves as the right wing of the hard pressed Western Front and neither the Admiralty nor their Flotilla Commander would restrict their operations. The Admiralty did carefully study the losses vs. successes and elected to gradually dissolve the Flanders Flotilla. The losses to the service of valuable and experienced personnel could not be easily replaced by any means. It would have been best to transfer the BII boats and experienced crews in the Spring of 1918 to the North Sea or Mediterranean respectively as the Commander of North Sea Submarines abandoned the use of the Dover Straits passage in February 1918. The latter order had been issued based on the experiences of KptLts. Rose and Saalwachter. Afterwards the average loss rate did not exceed two boats per month.

To counter the use of the Northern route, the English and Americans created a great mine barrage between the Orkneys and the Norwegian coast. It availed them little; only two boats are known to have been lost as a result of this field, "U156", under KptLt. Feldt, and "UB127" under Oblt. Scheffler. It cannot be denied the threat of this barrage was extremely unpleasant.

Germany's acceptance of Wilson's fourteen points in October 1918 brought about an end to war against trade. The idea of joint military operations with the Fleet, under Admiral Hipper, momentarily seemed possible and then was extinguished. (Admiral Scheer had been appointed Navy Supreme Commander.) The Fleet mutiny resulted in the recall of the B type boats which had assumed forward attack positions.

Submarine warfare did not bring about the forecast results and lessons were learned regarding its failure. The submarine war while unsuccessful exerted a considerable influence on the future of sea power and altered the conception of what constitutes sea power.

Fig 27. List of sinkings published in the White Books[210]

| War Period | Number of vessels over 100 tons | Ship tonnage in British registered tons |
|---|---|---|
| August 1914 to Sept 30,1915 | 431 | 791,705 |
| Oct.1,1915 to Apr. 30, 1916 | 359 | 898,794 |
| May 1, 1916 to Jan. 31, 1917 | 1,152 | 2,099,523 |
| Feb. 1, 1917 to Dec.31, 1917 | 2,566 | 5,753,751 |
| Jan. 1,1918 to end of war | 1,046 | 2,648,223 |
| Total I | 5,554 | 12,191,996 |

| War Period | Number of vessels over 100 tons | Ship tonnage in British registered tons |
|---|---|---|
| 1914 | 162 | 372,277 |
| 1915 | 726 | 1,438,173 |
| 1916 | 1,187 | 2,328,688 |
| 1917 | 2,734 | 6,350,362 |
| 1918 | 1,052 | 2,744,172 |
| Total II | 5,861 | 13,233,672 |
| Total I | 5,554 | 12,191,996 |
| Excess according to Lloyds | 307 | 1,041,676 |

Fig 28. Sinkings according to Lloyd's Register

210 Figs 27,28 data from Andreas Michelsen, Der U-Bootskrieg, 1914-1918. Leipzig. K. F. Koehler, 1925.

Printed in Great Britain
by Amazon

81312608R00181